INNOVATIVE SITE REMEDIATION TECHNOLOGY

STABILIZATION/ SOLIDIFICATION

One of an Eight-Volume Series

Edited by

William C. Anderson, P.E., DEE

Executive Director, American Academy of Environmental Engineers

1994

Prepared by WASTECH®, a multiorganization cooperative project managed by the American Academy of Environmental Engineers® with grant assistance from the U.S. Environmental Protection Agency, the U.S. Department of Defense, and the U.S. Department of Energy.

The following organizations participated in the preparation and review of this volume:

Air & Waste Management Association
P.O. Box 2861
Pittsburgh, PA 15230

American Society of Civil Engineers
345 East 47th Street
New York, NY 10017

American Academy of Environmental Engineers®
130 Holiday Court, Suite 100
Annapolis, MD 21401

American Society of Mechanical Engineers
345 East 47th Street
New York, NY 10017

American Institute of Chemical Engineers
345 East 47th Street
New York, NY 10017

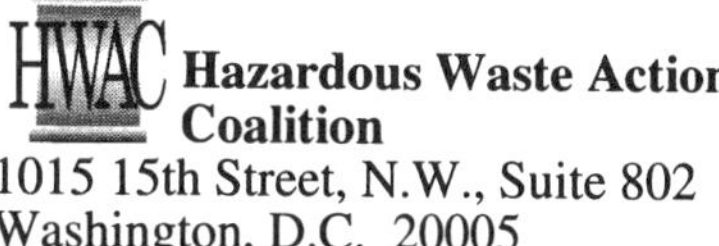

Hazardous Waste Action Coalition
1015 15th Street, N.W., Suite 802
Washington, D.C. 20005

Water Environment Federation
601 Wythe Street
Alexandria, VA 22314

Library of Congress Cataloging in Publication Data

Innovative site remediation technology/ edited by William C. Anderson
166p. 15.24 x 22.86cm.
Includes bibliographic references.
Contents: -- [3] Soil washing/soil flushing -- [4] Stabilization/solidification -- [6] Themal desorption.
1. Soil remediation. I. Anderson, William, C., 1943- .
II. American Academy of Environmental Engineers.
TD878.I55 1994 628.5'5--dc20 93-20786
ISBN 1-883767-03-2 (v. 3)
ISBN 1-883767-04-0 (v. 4)
ISBN 1-883767-06-7 (v. 6)

The material presented in this publication has been prepared in accordance with generally recognized engineering principles and practices and is for general information only. This information should not be used without first securing competent advice with respect to its suitability for any general or specific application.

The contents of this publication are not intended to be and should not be construed as a standard of the American Academy of Environmental Engineers or of any of the associated organizations mentioned in this publication and are not intended for use as a reference in purchase specifications, contracts, regulations, statutes, or any other legal document.

No reference made in this publication to any specific method, product, process, or service constitutes or implies an endorsement, recommendation, or warranty thereof by the American Academy of Environmental Engineers or any such associated organization.

Neither the American Academy of Environmental Engineers nor any of such associated organizations or authors makes any representation or warranty of any kind, whether express or implied, concerning the accuracy, suitability, or utility of any information published herein and neither the American Academy of Environmental Engineers nor any such associated organization or author shall be responsible for any errors, omissions, or damages arising out of use of this information.

Book design by Lori Imhoff

Printed in the United States of America

CONTRIBUTORS

This monograph was prepared under the supervision of the WASTECH® Steering Committee. The manuscript for the monograph was written by a task group of experts in stabilization/solidification and was, in turn, subjected to two peer reviews. One review was conducted under the auspices of the Steering Committee and the second by professional and technical organizations having substantial interest in the subject.

PRINCIPAL AUTHORS

Peter Colombo, *Task Group Chair*
Manager, Waste Management Research and Development
Brookhaven National Laboratory

Edwin Barth, P.E.
Environmental Engineer
Office of Research and Development
U.S. Environmental Protection Agency

Jim Buelt
Staff Engineer
Batelle Pacific Northwest Laboratory

Paul L. Bishop, Ph.D., P.E., DEE
William Thoms Professor
Department of Civil and Environmental Engineering
University of Cincinnati

Jesse R. Conner
Senior Research Scientist
Rust Remedial Services, Inc.
Clemson Technical Center

REVIEWERS

The panel that reviewed the monograph under the auspices of the Project Steering Committee was composed of:

Paul L. Busch, Ph.D., P.E., DEE, *Chair*
President
Malcolm Pirnie, Inc.

Roger Olsen, Ph.D.
Vice President
Camp Dresser and McKee

Bill Batchelor, Professor
Department of Civil Engineering
Texas A & M University

Ram Ramanujam, P.E.
Associate Waste Management Engineer
California EPA/DTSC

STEERING COMMITTEE

Frederick G. Pohland, Ph.D., P.E., DEE
Chair
Weidlein Professor of Environmental Engineering
University of Pittsburgh

William C. Anderson, P.E., DEE
Project Manager
Executive Director
American Academy of Environmental Engineers

Paul L. Busch, Ph.D., P.E., DEE
President and CEO
Malcolm Pirnie, Inc.
Representing, American Academy of Environmental Engineers

Richard A. Conway, P.E., DEE
Senior Corporate Fellow
Union Carbide Corporation
Chair, Environmental Engineering Committee
EPA Science Advisory Board

Timothy B. Holbrook, P.E.
District Engineering Manager
Groundwater Technology, Inc.
Representing, Air and Waste Management Association

Walter W. Kovalick, Jr., Ph.D.
Director, Technology Innovation Office
Office of Solid Waste and Emergency Response
U.S. Environmental Protection Agency

Joseph F. Lagnese, Jr., P.E., DEE
Private Consultant
Representing, Water Environment Federation

Peter B. Lederman, Ph.D., P.E., DEE, P.P.
Center for Env. Engineering & Science
New Jersey Institute of Technology
Representing, American Institute of Chemical Engineers

Raymond C. Loehr, Ph.D., P.E., DEE
H.M. Alharthy Centennial Chair and Professor
Civil Engineering Department
University of Texas

James A. Marsh
Office of Assistant Secretary of Defense for Environmental Technology

Timothy Oppelt
Director, Risk Reduction Engineering Laboratory
U.S. Environmental Protection Agency

George Pierce, Ph.D.
Editor in Chief
Journal of Microbiology
Manager, Bioremediation Technology Dev.
American Cyanamid Company
Representing the Society of Industrial Microbiology

H. Gerard Schwartz, Jr., Ph.D., P.E., DEE
Senior Vice President
Sverdrup
Representing, American Society of Civil Engineers

Claire H. Sink
Acting Director
Division of Technical Innovation
Office of Technical Integration
Environmental Education Development
U.S. Department of Energy

Peter W. Tunnicliffe, P.E., DEE
Senior Vice President
Camp Dresser & McKee, Incorporated
Representing, Hazardous Waste Action Coalition

Charles O. Velzy, P.E., DEE
Private Consultant
Representing, American Society of Mechanical Engineers

William A. Wallace
Vice President, Hazardous Waste Management
CH2M Hill
Representing, Hazardous Waste Action Coalition

Walter J. Weber, Jr., Ph.D., P.E., DEE
Earnest Boyce Distinguished Professor
University of Michigan

REVIEWING ORGANIZATIONS

The following organizations contributed to the monograph's review and acceptance by the professional community. The review process employed by each organization is described in its acceptance statement. Individual reviewers are, or are not, listed according to the instructions of each organization.

Air & Waste Management Association

The Air & Waste Management Association is a nonprofit technical and educational organization with more than 14,000 members in more than fifty countries. Founded in 1907, the Association provides a neutral forum where all viewpoints of an environmental management issue (technical, scientific, economic, social, political, and public health) receive equal consideration.

This worldwide network represents many disciplines: physical and social sciences, health and medicine, engineering, law, and management. The Association serves its membership by promoting environmental responsibility and providing technical and managerial leadership in the fields of air and waste management. Dedication to these objectives enables the Association to work towards its goal: a cleaner environment.

Qualified reviewers were recruited from the Waste Group of the Technical Council. It was determined that the monograph is technically sound and publication is endorsed.

The reviewers were:

James R. Donnelly
Director of Environmental Services and Technologies
Davy Environmental

Paul Lear
OHM Remediation Services, Corp.

American Institute of Chemical Engineers

The Environmental Division of the American Institute of Chemical Engineers has enlisted its members to review the monograph. Based on that review the Environmental Division endorses the publication of the monograph.

American Society of Civil Engineers

Qualified reviewers were recruited from the Environmental Engineering Division of ASCE and formed a Subcommittee on WASTECH®. The members of the Subcommittee have reviewed the monograph and have determined that it is acceptable for publication.

American Society of Mechanical Engineers

Founded in 1880, the American Society of Mechanical Engineers (ASME) is a nonprofit educational and technical organization, having at the date of publication of this document approximately 116,400 members, including 19,200 students. Members work in industry, government, academia, and consulting. The Society has thirty-seven technical divisions, four institutes, and three interdisciplinary programs which conduct more than thirty national and international conferences each year.

This document was reviewed by volunteer members of the Research Committee on Industrial and Municipal Waste, each with technical expertise and interest in the field covered by the document. Although, as indicated on the reverse of the title page of this document, neither ASME nor any of its Divisions or Committees endorses or recommends, or makes any representation or warranty with respect to, this document, those Divisions and Committees which conducted a review believe, based upon such review, that this document and the findings expressed are technically sound.

Hazardous Waste Action Coalition

The Hazardous Waste Action Coalition (HWAC) is an association dedicated to promoting an understanding of the state of the hazardous waste practice and related business issues. Our member firms are engineering and science firms that employ nearly 75,000 of this country's engineers, scientists, geologists, hydrogeologists, toxicologists, chemists, biologists, and others who solve hazardous waste problems as a professional service. HWAC is pleased to endorse the monograph as technically sound.

The lead reviewer was:

James D. Knauss, Ph.D.
Hatcher-Sayre, Incorporated

Water Environment Federation

The Water Environment Federation is a nonprofit, educational organization composed of member and affiliated associations throughout the world. Since 1928, the Federation has represented water quality specialists including engineers, scientists, government officials, industrial and municipal treatment plant operators, chemists, students, academic and equipment manufacturers, and distributors.

Qualified reviewers were recruited from the Federation's Hazardous Wastes Committee and from the general membership. A list of their names, titles, and business affiliations can be found listed below. It has been determined that the document is technically sound and publication is endorsed.

The reviewers were:

William Librizzi
Director of Government Programs
OHM Corporation

Linda Roberts Phipps, Ph.D., REM*
Process Chemist
RESOURCE CONSULTANTS, INC.

*WEF lead reviewer

ACKNOWLEDGMENTS

The WASTECH® project was conducted under a cooperative agreement between the American Academy of Environmental Engineers® and the Office of Solid Waste and Emergency Response,U.S. Environmental Protection Agency. The substantial assistance of the staff of the Technology Innovation Office was invaluable.

Financial support was provided by the U.S. Environmental Protection Agency, Department of Defense, Department of Energy, and the American Academy of Environmental Engineers®.

This multiorganization effort involving a large number of diverse professionals and substantial effort in coordinating meetings, facilitating communications, and editing and preparing multiple drafts was made possible by a dedicated staff provided by the American Academy of Environmental Engineers® consisting of:

Paul F. Peters
Assistant Project Manager & Managing Editor

Karen M. Tiemens
Editor

Susan C. Richards
Project Staff Assistant

J. Sammi Olmo
Project Administrative Manager

Yolanda Y. Moulden
Staff Assistant

I. Patricia Violette
Staff Assistant

TABLE OF CONTENTS

LIST OF TABLES

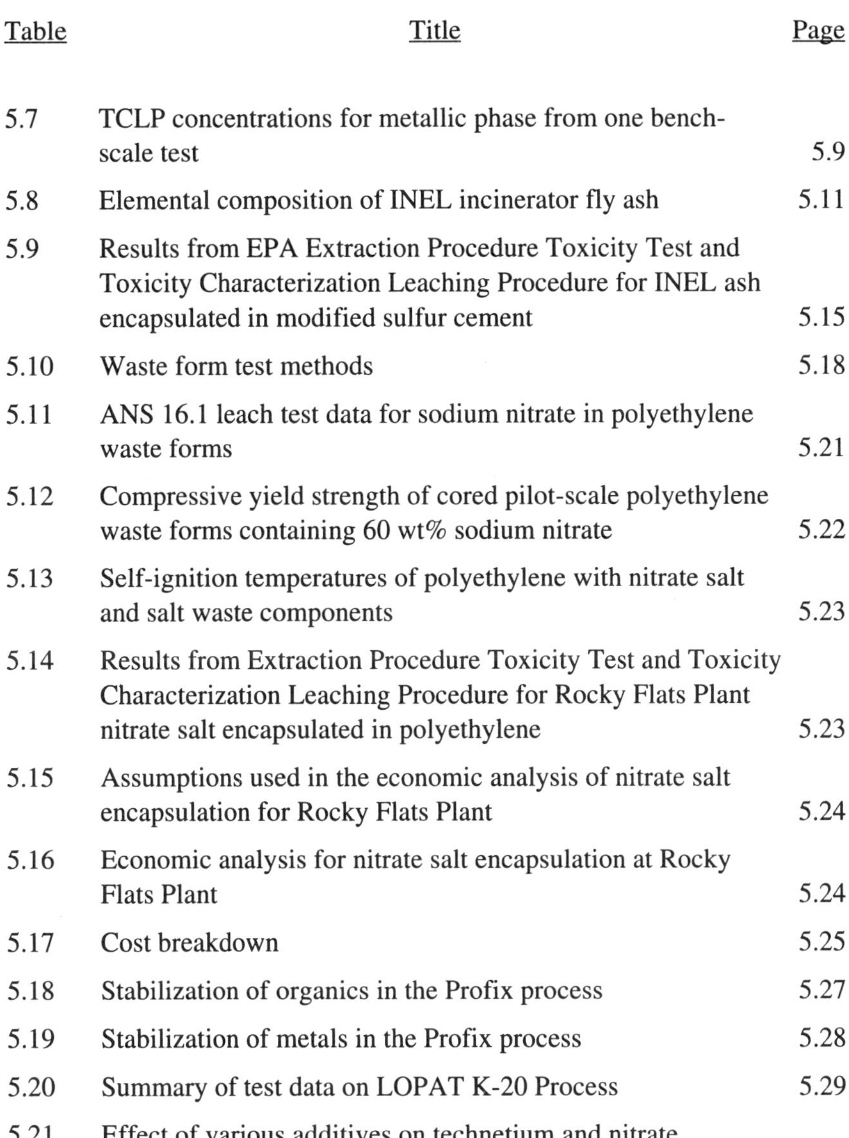

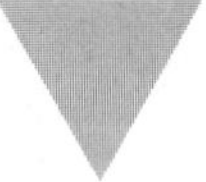

LIST OF FIGURES

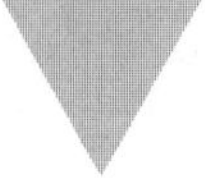

1 INTRODUCTION

This monograph on stabilization and solidification is one of a series of eight on innovative site and waste remediation technologies that is the culmination of a multiorganization effort involving more than 100 experts over a two-year period. It provides the experienced, practicing professional guidance on the application of innovative processes considered ready for full-scale application. Other monographs in this series address bioremediation, chemical treatment, chemical extraction, soil washing/soil flushing, thermal desorption, thermal destruction, and vacuum vapor extraction.

1.1 Stabilization and Solidification

Stabilization and Solidification are closely related in that both use chemical, physical, and thermal processes to detoxify a hazardous waste. But they are distinct technologies.

Stabilization refers to processes that reduce the risk posed by a waste by converting the contaminants into a less soluble, mobile, or toxic form. The physical nature of the waste is not necessarily changed.

Solidification refers to processes that encapsulate the waste in a monolithic solid of high-structural integrity. The encapsulation may be that of fine waste particles (microencapsulation) or of a large block or container of wastes (macroencapsulation). Solidification does not necessarily involve a chemical interaction between the waste and the solidifying reagents, but may mechanically bind the waste in the monolith. Contaminant migration is restricted by vastly decreasing the surface area exposed to leaching and/or by isolating the waste within an impervious capsule.

1.2 Development of the Monograph

1.2.1 Background

Acting upon its commitment to develop innovative treatment technologies for the remediation of hazardous waste sites and contaminated soils and groundwater, the U.S. Environmental Protection Agency (US EPA) established the Technology Innovation Office (TIO) in the Office of Solid Waste and Emergency Response in March, 1990. The mission assigned TIO was to foster greater use of innovative technologies.

In October of that same year, TIO, in conjunction with the National Advisory Council on Environmental Policy and Technology, convened a workshop for representatives of consulting engineering firms, professional societies, research organizations, and state agencies involved in site remediation. The workshop focused on defining the barriers that were impeding the application of innovative technologies in site remediation projects. One of the major impediments identified was the lack of reliable data on the performance, design parameters, and costs of innovative processes.

The need for reliable information led TIO to approach the American Academy of Environmental Engineers®. The Academy is a long-standing, multidisciplinary environmental engineering professional society with wide-ranging affiliations with the remediation and waste treatment professional communities. By June 1991, an agreement in principle (later formalized as a Cooperative Agreement) was reached. The Academy would manage a project to develop monographs describing the state of available innovative remediation technologies. Financial support would be provided by the EPA, U.S. Department of Defense (DOD), U.S. Department of Energy (DOE), and the Academy. The goal of both TIO and the Academy was to develop monographs providing reliable data that would be broadly recognized and accepted by the professional community, thereby, eliminating or, at least, minimizing this impediment to the use of innovative technologies.

The Academy's strategy for achieving the goal was founded on a multiorganization effort, WASTECH® (pronounced Waste Tech), which joined in partnership the Air and Waste Management Association, the American Institute of Chemical Engineers, the American Society of Civil Engineers, the American Society of Mechanical Engineers, the Hazardous

Waste Action Coalition, the Society for Industrial Microbiology, and the Water Environment Federation, together with the Academy, EPA, DOD, and DOE. A Steering Committee composed of highly respected representatives of these organizations having expertise in remediation technology formulated the specific project objectives and process for developing the monographs (see page iv for a listing of Steering Committee members).

By the end of 1991, the Steering Committee had organized the Project. Preparation of the monograph began in earnest in January, 1992.

1.2.2 Process

The Steering Committee decided upon the technologies, or technological areas, to be covered by each monograph, the monographs' general scope, and the process for their development and appointed a task group composed of five or more experts to write a manuscript for each monograph. The task groups were appointed with a view to balancing the interests of the groups principally concerned with the application of innovative site and waste remediation technologies — industry, consulting engineers, research, academe, and government (See page iii for a listing of members of the Stabilization/Solidification Task Group).

The Steering Committee called upon the task groups to examine and analyze all pertinent information available, within the Project's financial and time constraints. This included, but was not limited to, the comprehensive data base on remediation technologies compiled by US EPA, the store of information possessed by the task groups' members, that of other experts willing to voluntarily contribute their knowledge, and information supplied by process vendors.

To develop broad, consensus-based monographs, the Steering Committee prescribed a twofold peer review of the first drafts. One review was conducted by the Steering Committee itself, employing panels consisting of two members of the Committee supplemented by at least four other experts (See *Reviewers,* page iii, for the panel that reviewed this monograph). Simultaneous with the Steering Committee's review, each of the professional and technical organizations represented in the Project reviewed those monographs addressing technologies in which it has substantial interest and competence. Aided by a Symposium sponsored by the Academy in October 1992, persons having interest in the technologies were encouraged to participate in the organizations' review.

Comments resulting from both reviews were considered by the Task Group, appropriate adjustments were made, and a second draft published. The second draft was accepted by the Steering committee and participating organizations. The statements of the organizations that formally reviewed this monograph are presented under *Reviewing Organizations* on page v.

1.3 Purpose

The purpose of this monograph is to further the use of innovative stabilization and solidification site remediation and waste-processing technologies, i.e., technologies not commonly applied, where their use can provide better, more cost-effective performance than conventional methods. To this end, the monograph documents the current state of the technology for a number of innovative stabilization and solidification processes.

1.4 Objectives

The monograph's principal objective is to furnish guidance for experienced, practicing professionals and users' project managers. The monograph is intended, therefore, not to be prescriptive, but supportive. It is intended to aid experienced professionals in applying their judgment in deciding whether and how to apply the technologies addressed under the particular circumstances confronted.

In addition, the monograph is intended to inform regulatory agency personnel and the public about the conditions under which the processes it addresses are potentially applicable.

1.5 Scope

The monograph addresses innovative stabilization and solidification technologies that have been sufficiently developed so that they can be used in full-scale applications. It addresses all such technologies for which sufficient data were available to the Stabilization/Solidification Task Group to describe and explain the technology and assess its effectiveness, limitations, and potential applications. Laboratory- and pilot-scale technologies were addressed, as appropriate.

The monograph's primary focus is site remediation and waste treatment. To the extent the information provided can also be applied to production waste streams, it will provide the profession and users this additional benefit. The monograph considers all waste matrices to which stabilization and solidification processes can be reasonably applied, such as, soils, liquids, and sludges.

Application of site remediation and waste treatment technology is site-specific and involves consideration of a number of matters besides alternative technologies. Among them are the following that are addressed only to the extent essential to understand the applications and limitations of the technologies described:

- site investigations and assessments;
- planning, management, specifications, and procurement; and
- regulatory requirements.

1.6 Limitations

The information presented in this monograph has been prepared in accordance with generally recognized engineering principles and practices and is for general information only. This information should not be used without first securing competent advice with respect to its suitability for any general or specific application.

Readers are cautioned that the information presented is that which was generally available during the period when the monograph was prepared. Development of innovative site remediation and waste treatment technologies is ongoing. Accordingly, postpublication information may amplify, alter, or render obsolete the information about the processes addressed.

This monograph is not intended to be and should not be construed as a standard of any of the organizations associated with the WASTECH®

Project; nor does reference in this publication to any specific method, product, process, or service constitute or imply an endorsement, recommendation, or warranty thereof.

1.7 Organization

This monograph and others in the series are organized under a uniform outline intended to facilitate cross reference among them and comparison of the technologies they address. Chapter 2.0, Process Summary, provides an overview of all material presented. Chapter 3.0, Process Identification, provides comprehensive information on the processes addressed. Each process is analyzed in turn. The analysis includes, to the extent information and data are available, a description of the process (what it does and how it does it), its scientific basis, status of development, environmental effects, pre- and posttreatment requirements, health and safety considerations, design data, operational considerations, and comparative cost data. Also addressed are process-unique planning and management requirements and process variations.

Chapter 4.0, Potential Applications, Chapter 5.0, Process Evaluation, and Chapter 6.0, Limitations, provide syntheses of available information and informed judgments on the processes. Each of these chapters addresses the processes in the same order as they are described in Chapter 3.0. Technology Prognosis, Chapter 7.0, identifies aspects of each of the processes needing further research and demonstration before full-scale application can be considered.

2
PROCESS SUMMARY[1]

2.1 Process Identification and Description

Innovations in stabilization and solidification technology abound. This monograph addresses eight distinct innovative processes or groups of processes. Most of the innovations are modifications of proven processes and are directed to encapsulating or immobilizing the harmful constituents and involve excavation and processing of the waste or contaminated soil.

2.1.1 Sorption and Surfactant Processes

Sorption processes are based on a contaminant being attracted and retained on a sorbent. Interactions between inorganic heavy metals and ion exchange media, clay, humic material, fly ashes, activated carbon, etc., are well documented in the waste management literature.

In general, hydrophobic organic material is not compatible with inorganic material such as cement. By substituting quaternary ammonium ions for group IA and IIA metal ions in clays, however, the interplaner distance between aluminum and silica can be increased allowing clays to sorb organic compounds. The resulting clay/organic interactions vary from weak to strong and are solidified by the addition of cement.

In simple terms, surfactants are manufactured to have different compatibilities on each end of the molecule, allowing waste material to be sorbed on one end, while the other end is compatible with inorganic cement. Surfactants can be used also to disperse organic waste material in an aqueous phase and then combine with cement for solidification.

---

1. This chapter is a summary of Chapters 3.0 through 7.0. Sources are cited, where appropriate, in those chapters — Ed.

These processes, typically protected by patents, often combine the use of both sorbent additives and surfactants. Examples of companies that have provided field-scale services using these processes include Silicate Technology Corporation (STC), International Waste Technologies (IWT), Hazcon, Soliditech, and Wastech.

2.1.2 Emulsified Asphalt

Asphalt emulsions are very fine droplets of asphalt dispersed in water that are stabilized by chemical emulsifying agents. Emulsified asphalt is commercially available either as an anionic or a cationic emulsion. For the highest efficiency, the most appropriate emulsified asphalt for the waste to be treated must be selected.

The process involves adding emulsified asphalts having the appropriate charge to hydrophilic liquid or semiliquid wastes at ambient temperature. After mixing, the emulsion breaks, the water in the waste is released, and the organic phase forms a continuous matrix of hydrophobic asphalt around the waste solids. In some cases, additional neutralizing agents, such as lime or gypsum, may be required. After given sufficient time to set and cure, the resulting solid asphalt has the waste uniformly distributed throughout it and is impermeable to water.

2.1.3 Bituminization

In the bituminization process, wastes are embedded in molten bitumen and encapsulated when the bitumen cools. The process combines heated bitumen and a concentrate of the waste material, usually in slurry form, in a heated extruder containing screws that mix the bitumen and waste. Water is evaporated from the mixture to about 0.5% moisture. The final product is a homogenous mixture of extruded solids and bitumen.

2.1.4 Vitrification

Vitrification processes are solidification processes that employ heat to melt and convert waste materials into glass or other glass and crystalline products. Waste materials, such as heavy metals and radionuclides, are actually incorporated into the glass structure which is, generally, a relatively strong, durable material that is resistant to leaching.

Vitrification involves glass formers, stabilizers, and fluxes. Silica is the principal glass former and provides the basic matrix of the vitrified product. In waste management, silicates and borosilicates are used. Fluxes, primarily sodium oxide that generally exist within the waste material, reduce the melting temperature and, when molten, increase electrical conductivity, thereby enhancing process efficiency. Stabilizers are employed to increase the durability of the glass and to decrease the conductivity of the molten pool.

The high-temperatures employed, 1,200°C or higher, cause the glass formers, fluxes, stabilizers, and wastes to melt together to form a glass matrix. The high temperatures also destroy any organic constituents with very few byproducts, which are treated with an offgas treatment system that generally accompanies vitrification processes. There are three basic technology variations that differ in the kind of heating employed – electrical processes, thermal processes, and plasma systems.

2.1.4.1 Electrical Processes

Electrical vitrification processes employ joule heating in which electrical energy is imparted to glass to create a molten pool. Convective currents, generated within the molten pool, enhance mixing, creating a more homogenous product. Further, cleaning of the process offgases is simpler than it would be if fossil fuels were used, since no excess air for combustion of fuels is required.

Electrical processes are generally applied above ground in refractory-lined or water-cooled melters of either horizontal or vertical design using electrodes or radiant heating. Waste materials, which can be slurries, wet or dry solids, or combustible material, are mixed with glass formers and conveyed to the molten pool in the melter. Offgases are cleaned and the glass formed withdrawn continuously or intermittently through a bottom drain or overflow weir. Processing rates range from 80 to 220 kg/hr (175 to 485 lb/hr), but may be as high as 3,600 kg/hr (9,900 lb/hr) for municipal waste vitrification.

Electrical processes are also applied in situ by inserting 4 graphite electrodes a few centimeters into the contaminated soil, laying on a mixture of ground glass frit and graphite flakes, and then applying an electrical current. Once molten, the soil conducts electrical current causing the molten soil to grow outward and downward. Blocks, up to 12 m (13 yd) on each side, can be created in this batch-type application. Processing rates range from 2.7 to 4.5 tonne/hr (3 to 5 ton/hr).

2.1.4.2 Thermal Processes

Thermal processes require an external heat source, typically fossil fuels, to heat the waste constituents, glass formers, and fluxes via convective and radiative heat. The typical reactor is a refractory-lined rotary kiln.

2.1.4.3 Plasma

Plasma systems employ temperatures up to 10,000°F, created by an electrical discharge in an arc or torch through a gas. The resulting plasma destroys organic constituents of the waste and liquefies the remainder through radiative heat transfer. The process is operated on a batch basis and requires treatment of the offgas. Compared with thermal processes, plasma systems have much faster rates of heat transfer and can process waste up to 0.45 tonne/hr (0.5 ton/hr).

2.1.5 Modified Sulfur Cement

Modified sulfur cement is a commercially-available thermoplastic material. It is easily melted (127° to 149°C (260° to 300°F)) and then mixed with the waste to form a homogenous molten slurry which is discharged to suitable containers for cooling, storage, and disposal. A variety of common mixing devices, such as, paddle mixers and pug mills, can be used. The relatively low temperatures used limit emissions of sulfur dioxide and hydrogen sulfide to allowable threshold values.

2.1.6 Polyethylene Extrusion Process

The polyethylene extrusion process involves the mixing of polyethylene binders and dry waste materials using a heated cylinder containing a mixing/transport screw. The heated, homogeneous mixture exits the cylinder through an output die into a mold, where it cools and solidifies. Polyethylene's properties produce a very stable, solidified product. The process has been tested on nitrate salt wastes at plant-scale, establishing its viability, and on various other wastes at the bench- and pilot-scale.

2.1.7 Inorganic, Cementitious Technologies of the Siliceous Category

Cementitious stabilization is applicable to a wide range of industrial wastes and results in very stable products. It is a well-established process;

the most common are lime-fly ash and portland cement-sodium silicate systems. Innovations include modifications of established systems and the use of other reagents.

2.1.7.1 Soluble Silicate Processes

Soluble silicates are applied either as accelerators or retarders in cementitious systems to reduce leachability. A variety of conventional patented processes employ this technology. Innovations in chemistry of the use soluble silicates include enviroGuard™, enviroGuard Plus™, ProFix™, and Lopat. There are many other patented processes employing this technology, although none is commercialized.

2.1.7.2 Slag Processes

Slags, themselves waste products, have been used for several years in waste treatment. Slags, either alone or with cementitious materials, are mixed with waste slurries to enhance settling and compaction, after which the supernatant is drawn off leaving a stable residue. Typically, slag processes are applicable to dilute wastes, require large areas for settling and compaction ponds, and can be conducted over long periods.

The Oak Ridge National Laboratory (ORNL) Process employs blast-furnace slag in combination with portland cement and fly ash for stabilization of radioactive wastes. SoliRoc™ is a complete hazardous waste treatment system that employs a complex pretreatment process using blast-furnace slag, precipitation using process-generated soluble silicates, sludge dewatering, and solidification with portland cement.

Another innovation employs portland cement and slag to stabilize hexavalent chromium. Laboratory leachability tests confirm that the Cr^{+6} could be stabilized to within US EPA Toxicity Characteristic Leaching Procedure (TCLP) limits.

2.1.7.3 Lime

Another process uses lime that has been pretreated with additives causing it to be hydrophobic. Patented processes offering this technology are Separation and Recovery Systems (SRS)/EIF and DCR/Boelsing/Sound Environmental Services, Inc. According to the patent, these processes enable lime to sorb organics and encapsulate them within an insoluble calcium carbonate.

2.1.7.4 Inorganic Polymers

In soluble silicate processes, some polymerization may occur. The "Geopolymer" process, however, appears to form inorganic polymers based on silicon and aluminum through chemical reaction. The geopolymers act as binders when mixed with wastes, producing a high-viscosity mixture that can be molded. After curing, the resulting product is characterized by high-strength, hardness, and resistance to chemical attack, especially attack by acids.

2.1.8 Soluble Phosphates

The process involves the addition of various forms of phosphate and alkali for control of pH as well as for formation of complex metal molecules of low-solubility to immobilize (insolubilize) the metals over a wide pH range. Unlike most other stabilization processes, soluble phosphate processes do not convert the waste into a hardened, monolithic mass. Rather, the waste remains free-flowing and increases little in volume. Process chemistry is detailed in Chapter 3.0.

Soluble phosphates and lime have been used commercially to stabilize fly ash. Expanding upon this proven data base, Wheelabrator Environmental Systems developed the patented WES-PHix process for immobilizing lead and cadmium in other ash residues and waste streams. This totally-enclosed, in-line system has been permitted by several states. Because it is totally enclosed, health or safety problems for workers are minimal. The resulting product, which reduces lead and cadmium leaching below Toxicity Characteristic Leaching Procedure (TCLP) limits, is in particulate form. Mix formulations must be determined on a site-specific basis, but any convenient source of water soluble phosphates may be used. Some guidance on chemical-waste ratios are provided in Chapter 3.0.

2.2 ***Potential Applications***

Stabilization and solidification processes can be applied to a wide range of wastes. Since all the processes depend upon chemical and physical reactions of varying complexity between the wastes and the applied fixation

Table 2.1
Potential Applications

Stabilization/ Solidification Process	Kinds of Wastes
Sorbents and Surfactants	Oily wastes, industrial sludges and contaminated soils (containing inorganics with low concentrations of organics), acid mine leachate and tailings, and radioactive liquid scintillation fluids
Emulsified Asphalt	Contaminated foundry sands and sand-blasting grit; wastes from paint removal, metal finishing, and electroplating; petroleum-contaminated soils
Soluble Phosphates	Refuse-to-energy plant and medical wastes ash, insulation wastes, metals-smelting dusts, contaminated soils, and metal-contaminated sludges
Bituminization	Low- and medium-level radioactive waste solutions, and metal plating sludges
Electrical Process Vitrification	High-level radioactive wastes, municipal solid waste ash, medical wastes, and contaminated soils and sludges
In Situ Vitrification	Soils or sludges contaminated with radioactive, metallic, or organic wastes; particularly suited for large quantities of in-place or stock-piled soils or sludges
Thermal Vitrification	Contaminated soils, incinerator fly ash, organic wastes, and nonvolatile metallic wastes
Plasma Vitrification	Small, concentrated quantities of slurries and contaminated solid materials, such as, metals, glass, and filter elements
Modified Sulfur Cement Process	Predried particulate wastes, such as, incinerator ash, contaminated soils, sludges, metals, and mill tailings
Polyethylene Extrusion	Primary and secondary waste streams produced by waste management and restoration activities, nitrate salt wastes, sludges, ion-exchange resins, incinerator ash, and scrubber blow-down solution
Soluble Silicates (Patented Processes)	(Specific process variation dependent) metal-containing wastes, auto shredder fluff, incinerator-bottom ash, and contaminated debris
Soluble Silicates (Slags)	Metal-refining wastes, metal-finishing wastes, and metal-bearing sludges
Soluble Silicates (Lime)	Steel pickle liquor, ferric chloride etching waste, oil waste, hydrocarbon waste, incinerator ash, petroleum sludge, phosphoric acid residue, oily wastes, and tars
Soluble Silicates (Inorganic Polymers)	(Specific process variation dependent) metal-containing wastes, auto shredder fluff, incinerator-bottom ash, and contaminated debris

agents, care is required in order to select the most appropriate process or system. Table 2.1 lists the most common kinds of wastes to which each process described in this monograph is applicable.

2.3 *Process Evaluation*

Available performance and cost data for each of the stabilization/solidification processes examined in this monograph are compiled and summarized in Chapter 3.0. These processes depend upon physical and chemical reactions of varying complexity between the wastes and the fixation agents. Accordingly, performance data provided are limited, for the most part, to those combinations that have achieved the greatest success.

2.3.1 Sorption and Surfactant Processes

Sorption and surfactant processes have evolved from bench-scale to full-scale implementation. When the various binders are correctly matched with waste materials, these processes appear to be able to immobilize many organic constituents present at low concentrations, at least temporarily. Generally, inorganic solids (such as pozzolanic material) are weak sorbents for organic materials. Quaternary ammonium compounds that are added to clays have been extensively studied for sorbing organics. Available cost data are site-specific and generally do not permit cost comparison of vendors.

2.3.2 Emulsified Asphalt

The emulsified asphalt process has proven to be effective in solidifying liquid wastes, reducing solidified waste permeability, and improving the effectiveness of stabilization/solidification processes using portland cement and soluble silicates. The cost to treat wastes using this process generally ranges from $90 to $110/tonne ($80 to $100/ton), but can be as high as $165/tonne ($150/ton).

2.3.3 Bituminization

Bituminization has proven to be effective in treating low-level radioactive wastes. While radionuclides are effectively controlled, there are no data for heavy metals or organics.

2.3.4 Vitrification

Each of the vitrification processes reported has been extensively studied. Generally, all vitrification processes are very effective in treating radwastes, heavy metals, and in many cases organics, with total destruction and removal efficiencies ranging from 99.99% to 99.99999% (seven nines). Some metals are volatilized and must be captured with offgas treatment systems.

2.3.5 Modified Sulfur Cement Process

The majority of application studies of the modified sulfur cement process have focused on incinerator fly ash. Modified sulfur cement alone is a relatively brittle material, but in combination with wastes (values are not highly dependent on waste loading), its compressive strength is more than doubled. To meet TCLP criteria, modified sulfur cement may be augmented (sodium sulfide has proven effective in a mix containing, by weight, 40% waste, 53% modified sulfur cement, and 7% sodium sulfide).

2.3.6 Polyethylene Extrusion Process

Polyethylene waste forms containing various radioactive and hazardous wastes have been extensively tested. Most of these tests were conducted with sodium nitrate waste and proved the process can produce a product that is nonhazardous by all applicable criteria. Polyethylene encapsulation offers potentially significant cost savings over cement-based encapsulation when all relevant cost factors are considered.

2.3.7 Inorganic, Cementitious Technologies of the Siliceous Category

With the exception of geopolymers, these processes are well-established and employ a range of commercially-available reagents. They have proven their ability to eliminate the hazardous characteristics of most wastes. However, it is important to match the particular process to the waste stream. The cost effectiveness of most of these processes depend entirely upon the reagent employed and its availability at a particular site, although a royalty may be exacted for some of the patented processes, such as Lopat or EnviroGuard™.

Inorganic geopolymers have not been used commercially for the stabilization of wastes, but reportedly have been used in Europe and Canada for construction purposes. Tests on various waste streams have established the effectiveness of such polymers to eliminating hazardous characteristics. However, geopolymerization takes a relatively long period, up to 21 days.

2.3.8 Soluble Phosphates

Evaluation of the WES-PHix process has shown it to be far more effective than portland cement treatment for control of zinc, copper, cromium, and lead and similar in performance for control of cadmium. Soluble phosphates reduced lead in leachates by a factor of 6 in bottom ash and by a factor of 900 in leachates from air pollution control residues compared with treatment by portland cement alone.

2.4 *Limitations*

While the eight general categories of stabilization/solidification processes, and their variations, are distinct in many regards, most have remarkably similar limitations. Following is a summary of limitations:

- waste composition – each of the processes is tailored to one particular kind of waste or to a family of wastes, there is not one best process for all waste types;
- reagent availability – the cost-effectiveness of those processes that employ reagents is substantially affected by reagent availability at the site;
- moisture content – for some processes, there can be too much moisture in the waste for the process to perform as intended or to be cost-effective. For others, there can be too little moisture; and,
- risk of release – for almost all processes, except vitrification, there remains the risk, to varying degrees, that the stabilized/ solidified waste matrix will break down over time, leading to release of harmful constituents into the environment.

2.5 *Technology Prognosis*

Each of the processes have proven their suitability for at least one kind of waste. Additional research will enable application of each to be expanded to other wastes. Some of the more promising areas of development or key information requirements of each process are summarized, below:

- Sorption/Surfactant – development of additional natural or synthetic binders; also needed is information on saturation limits and the degree of sorption or degradation under realistic disposal conditions;
- Emulsified Asphalt – development to apply to wastes containing metals and nonpetroleum organics;
- Bituminization – development of more durable waste/bitumen mixtures;
- Vitrification – refinement of the technology to spread application beyond contaminated soil applications, and improve process efficiency (e.g., process rate, cost, energy utilization);
- Modified Sulfur Cement – determination of optimum waste loadings and application to sludges and contaminated soils;
- Polyethylene Extrusion – full-scale demonstration of process reliability and determination of maximum heavy metals loadings;
- Soluble Silicates – more information is needed on each of the various technologies involved to broaden their application and/or enhance their performance and reliability; and
- Soluble Phosphates – development to apply to metals-bearing wastes.

PROCESS IDENTIFICATION AND DESCRIPTION

3.1 *Sorption and Surfactant Processes*

3.1.1 Description

Sorption processes are based on a contaminant being attracted to and retained on a sorbent. Various attraction forces act to partition the contaminant out of the aqueous phase and onto the surface of the sorbent. The interaction forces are of various degrees, such as hydrogen bonding, ion exchange reactions, and may include van der Waals or electrostatic forces.

Surfactants are manufactured so as to have different chemical compatibilities on each end of the molecule. In simple terms, this configuration allows organic waste material to be sorbed on one end, while the other end is compatible with water mixed with the inorganic cement.

3.1.2 Scientific Basis

In general, hydrophobic organic material is not compatible with cement. Therefore, added organic sorbents can combine with organic waste material before being solidified in cement. For example, quaternary ammonium ions (R_4N^+) can be substituted for group IA and IIA metal ions (Li^+, Na^+, K^+, Mg^{2+}, Ca^{2+}, and Ba^{2+}) yielding clays that have both organic and inorganic reactive properties, enabling the clay to also sorb organic compounds. These substitutions sterically increase interplanar distance (allowing compound insertion) and create an organic stationary phase with polarity compatible with the organic waste. Chemical interactions that range in strength retain the organic compound near the clay surface. Cement is added for solidification to reduce available surface area accessible to leaching.

With the use of a surfactant to lower surface tension, organic waste material can be dispersed with water in which the continuous phase is aqueous. This suspension can then be mixed with cement for solidification.

3.1.3 Status and Development

There are several patents registered covering a variety of processes. Some of them cover a combination of sorbent additives or the use of surfactants for dispersion.

Listed below are some of the commercial vendors who have provided field-scale services:

- Silicate Technology Corporation (STC);
- International Waste Technologies (IWT);
- Hazcon;
- Soliditech; and
- Wastech.

Stabilization/solidification data relating to some of these vendors' processes are provided in Section 4.1.

3.1.4 Pretreatment Requirements

For some processes, pretreatment would include mixing the contaminant and sorbent before adding cement. As an example, surfactants are first utilized to emulsify organic waste and then this dispersion is mixed with cement.

3.1.5 Health and Safety Considerations

Volatile organic compounds that are not instantaneously sorbed might be released during a mixing operation because of the exothermic cement/water reaction and increased surface area exposure. Since stabilized organic material is normally disposed of in the soil, there will always remain a threat to groundwater because of possible contaminant diffusion from the treated matrix.

Some sorbents, such as quaternary ammonium compounds, may have detrimental environmental effects if released into surface water.

3.1.6 Operational Considerations

The amount of sorption chemicals that is used to stabilize a waste, plus the amount of cement needed to solidify a waste, compared to the waste amount is normally referred to as a binder-to-waste ratio. Contaminant sorption is a function of several factors, such as pH, concentration of contaminants, selectivity of sorbent surface, and kind of aqueous disposal environment. Bench-scale column release testing is useful in determining the amount of sorbent needed.

Processing capacity is largely dependent on the time required to get adequate mixing of the sorbent, waste, and cement.

Among the more important operating parameters is the degree of mixing between the contaminant and sorbent and between the sorbent and cement.

Employing an additional reagent that is natural to a cement-based process should increase cost very little relative to cement alone because the materials are generally available at low cost. Furthermore, sorption may reduce the amount of cement necessary to minimize leaching. Synthetic compounds, however, such as quaternary ammonium compounds, are relatively expensive compared with other natural sorbents, such as sawdust, humic materials, etc.

The designer of an organic sorbent stabilization process must know the degree of interaction between the contaminants and sorbent under the conditions existing at the time of stabilization/ solidification.

There is practically an unlimited number of variations of sorption processes, since each vendor can use more than one sorbent or combination of sorbents, with or without a surfactant.

3.2 *Emulsified Asphalt*

3.2.1 Description

Asphalt-based solidification processes have been investigated since the 1950s for encapsulation of nuclear wastes. Little research was done until recently, however, on use of asphalt for treatment of hazardous wastes or contaminated soils.

One promising technology for fixation of hazardous wastes and contaminated soils is based on the use of emulsified asphalt. Asphalt emulsions are intimate mixtures of asphalt and water. Very fine droplets of asphalt are dispersed in water to create emulsions. Chemical emulsifying agents, such as detergents, are then added to make the product stable. They form a protective film around the emulsified asphalt droplets and carry an electric charge that causes the droplets to repel one another.

This process operates at ambient temperature, thereby eliminating the problems associated with hot asphalt mix processes, namely, volatilization of contaminants and high energy costs. Hydrophilic liquid and semi-liquid wastes can be treated with emulsified asphalt and, possibly, with other solidification additives, rendering the wastes hydrophobic and impermeable to water. The process can be used with high-solids wastes, such as sludges, with contaminated soils, or with contaminated liquids.

3.2.2 Scientific Basis

In emulsified asphalt processes, the waste is mixed with the suspension of emulsified asphalt particles at ambient temperature in an amount sufficient to react with counter-ions in the waste and coalesce into a hydrophobic mass. After mixing, the mixture will destabilize chemically, causing the asphalt emulsions to "break" and solidify. The water in the waste is released, forming a biphasic mixture. The organic phase forms a continuous matrix of hydrophobic asphalt around the waste solids (Conner 1990). The released water may need further treatment.

The mixture is then allowed sufficient time to set and cure. The resultant end product is a solid material varying in consistency from a rock-like solid to a friable material. Addition of the emulsified asphalt to the waste material increases the hydrophobicity and decreases the water permeability of the end product. Thus, the asphalt-treated waste material is no longer accessible to water-based leachates, and the end product is highly stable and substantially impervious to aqueous leaching of waste constituents.

3.2.3 Information Necessary to Employ Process

The ionic charge of the waste counter-ions is determined empirically by adding cationic and anionic asphalt emulsions to small samples of waste and observing which mixtures coalesce. Alternatively, the charge on waste particles can be determined by using an electrokinetic zeta potential measuring device such as a "Zeta Meter." An appropriate suspension of emulsified asphalt particles having an opposite particle charge to the ionic charge of the waste contaminant is then selected.

3.2.4 Operational Considerations

Emulsified asphalt is commercially available either as a cationic or an anionic emulsion. The asphalt is added to the waste in quantities ranging from 0.1% to about 40% by weight. After mixing the asphalt emulsion with the waste material, solidification is promoted by neutralizing the surface charge on the emulsified asphalt particles with the waste counter-ions. This reduces the charge repulsion between particles and allows the particles to coalesce into a hydrophobic mass, leaving higher quality water behind. In some cases, it may be necessary to add additional neutralizing agents, such as lime (Ca^{2+}) for neutralizing anionic asphalt emulsions or gypsum (SO_4^{2-}) for neutralization of cationic ones. The resulting solid asphalt contains the waste material uniformly distributed throughout it. The hydrophobic properties cause it to have very low permeability to water.

3.3 Bituminization

3.3.1 Description

Bitumen, or asphalt, is a relatively leach-resistant, thermoplastic substance. Contaminated residues can be embedded in molten bitumen which, when it cools, encapsulates the residues. Heated bitumen is extruded from a storage tank at approximately 10 to 50 L/hr (2.6 to 13 gal/hr) into an extruder. Simultaneously, the concentrate of contaminated material, usually in slurry form, is pumped into the extruder. Processing rates as high as 300 kg/hr (660 lb/hr) are available. Screws inside the extruder mix the material until it is conveyed to a discharge pipe. Water is evaporated from the mixture until the residual moisture content is about 0.5% by weight (Simon and Botzem 1984). The final product, a homogeneous mixture of extruded solids and bitumen, is then discharged from the extruder into a 55-gallon drum.

The process achieves a volume reduction for slurry materials of about 60%. Generally 1 m^3 (1.3 yd^3) of concentrated slurry, when mixed with bitumen, can be converted to about 0.4 m^3 (0.5 yd^3) of final product. This value is highly dependent, however, on the kind of waste. Westsik (1984) reported volume changes, varying with waste stream characteristics; volume reductions of 2.4:1 of waste:bitumen product were achieved, i.e., the volume of original waste slurry before treatment was 2.4 times the volume of the final bitumen product. But in some cases, volume increases of 0.5:1 resulted.

3.3.2 Operational Considerations

Processing rates as high as 300 kg/hr (660 lb/hr) can be achieved at existing process scale (Simon and Botzem 1984). The amount of electrical power required to maintain temperature within the extruder is approximately 20 kw. A bituminization extruder of this capacity is approximately 6 m (20 ft) in length by 1.5 m (5 ft) in width by 2 m (7 ft) in height. The overall dimensions of the bituminization plant would be approximately 15 m (50 ft) by 10 m (33 ft) by 4 m (13 ft). Figure 3.1 (on page 3.7) provides a schematic of a typical bituminization plant.

3.4 Vitrification

3.4.1 Description

Vitrification is a class of thermal treatment processes that convert waste materials into a glass or glass and crystalline product. Waste materials, such as heavy metals or radionuclides, are not simply encapsulated, but are actually incorporated in their oxide form into the structure of the glass product. Metallic components that may coexist with the waste materials separate into a different phase that can be tapped separately in many vitrification processes. This process results in a relatively strong and durable material that is capable of lasting for thousands of years.

Figure 3.1
Simplified Diagram of a Bituminization Plant

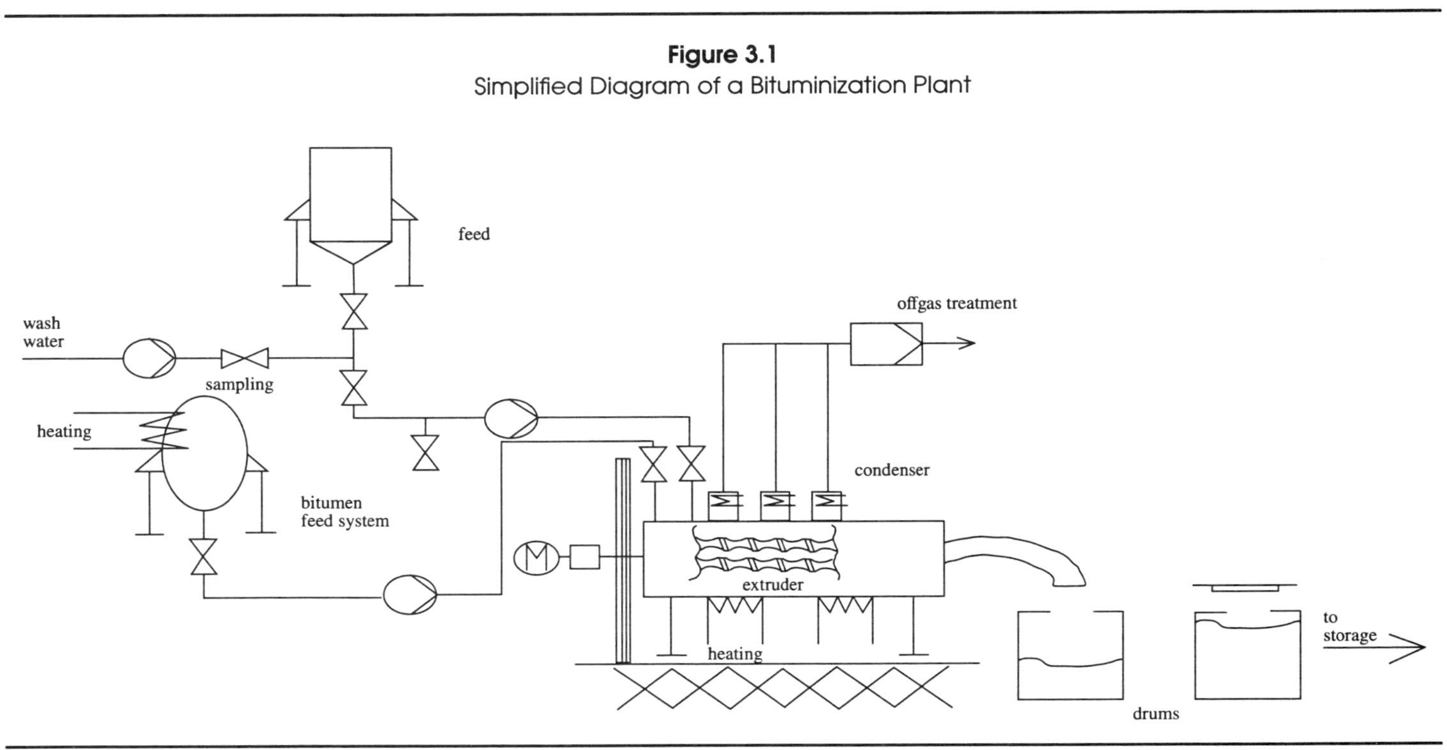

3.4.2 Scientific Basis

Vitrification products are commonly composed of oxides that can be classified into one of three primary groups: (1) glass formers, (2) stabilizers, and (3) fluxes (Tooley 1974).

For most glasses, the principal glass-former constituent is silica (SiO_2) (US EPA 1992b). Glass formers provide the basic silicon-oxygen tetrahedron matrix of vitrified products. Other glass formers such as phosphates and borates exist, but waste remediation technologies are generally limited to silicate and borosilicate glasses.

Fluxes reduce the melting temperature and viscosity of the oxides within the glass matrix, as well as increase the electrical conductivity of glasses when molten. Therefore, fluxes become important to the processibility of the vitrified product, especially for electrically heated vitrification processes. The primary flux constituent is sodium oxide (Na_2O), but the flux can also contain potassium oxide (K_2O), lithium oxide, and other alkaline oxide chemicals. Fluxes can be added in more common chemical forms, such as soda ash (Na_2CO_3), and be allowed to decompose to their oxide form during processing. They can also be added in chemical forms that include glass formers, such as sodium silicate (Na_2SiO_3).

Stabilizers are most responsible for increasing the durability and decreasing the electrical conductivity of molten glasses. The primary constituents of stabilizers are calcium oxides (CaO) and other alkaline earth oxides. Alumina (Al_2O_3) can also act as a modifier. Stanek (1977) shows the relative impact of stabilizers on electrical conductivity in relation to other constituents. The relative influence is shown in table 3.1 (on page 3.9).

When heated, the constituents decompose to their oxide form, fuse, and melt together to form a glass matrix. For example, in a simple system, the fusion of SiO_2 and Na_2CO_3 occurs when Na_2CO_3 decomposes to form Na_2O and CO_2, and Na_2O fluxes with SiO_2 to form a melt at less than the melting temperature of SiO_2. The sodium interferes with some of the silicon-oxygen-silicon bonds, thus disrupting the polymerization of units of tetrahedral structure of SiO_2 and reducing the melting temperature and viscosity of the mixture. Similarly, the addition of a stabilizer, such as CaO, can lower the melting point, yet maintain the durability of the glass (US EPA 1992b).

In most vitrification processes, the glass product can be designed with specific compositions in order to obtain specific physical and chemical

Table 3.1
Influence of Constituents on Electrical Conductivity

Influence	Oxide
Greatly Decrease	CaO
Slightly Decreased	B_2O_3
Neutral	BaO Fe_2O_3 PbO MgO ZnO SiO_2
Slightly Increased	Al_2O_3
Greatly Increased	K_2O
Extremely Increased	Na_2O

Note: Degree of influence of individual oxides on the electrical conductivity of glass, CaO effecting the most significant reduction and Na_2O, the most significant increase.
(Stanek 1977)

properties. Waste constituents, such as lead or chromium, also enter into the formation of the glass product. These constituents can act as glass formers, replacing silica in the structure or acting as stabilizers and fluxes and interfering with the bonds of silicon and oxygen ions. Consequently, the waste constituents become part of the glass, making them more resistant to chemical attack than if they were merely encapsulated within the glass.

The high temperature of the vitrification process also destroys organic constituents. Most vitrification processes operate at 1,200°C (2,200°F) or higher, a temperature capable of destroying any organic contaminant. The organic contaminants decompose into carbon dioxide (CO_2), water (H_2O), and hydrogen chloride (HCl) when interacted with oxygen as the gases exit the vitrification process.

3.4.3 Technology Variations

There are three basic variations in the technology of vitrification processes – electrical, thermal, and plasma.

3.4.3.1 Electrical Processes

Many vitrification processes operate on the principle of joule heating. Joule heating is the mechanism by which electrical energy is imparted to the glass to generate heat. Normally, glasses are not electrically conductive, but when in the molten state, the alkaline elements within the glass ionize and become mobile, allowing them to transmit an electrical charge. This electrical charge is usually transmitted in a 60 Hz alternating current. An electrical voltage is applied to electrodes at either end of the melting cavity. The joule heating effect causes the alternating current to flow between the electrodes, thereby generating the heat necessary for continued melting of material to be vitrified.

As more heat is generated within the molten pool, the molten glass becomes less viscous and more electrically conductive. Convective currents within the molten pool keep the material well mixed. It is this combined effect of convective flow and joule heating that makes the electrical melting process so effective.

Electrical processes generally operate at electrical voltages from approximately 100 volts up to thousands of volts. Consequently, operational procedures and engineering safety features must be designed to avoid the potential for electrical shock to the operators.

Electrical processes have a distinct advantage in terms of gaseous effluent generation. Because no fossil fuel is required to generate the heat, the gases that are generated are limited primarily to water vapor, from the drying of the waste materials, and decomposition gases, such as CO_2, that form during the vitrification process. Residues from combustion of fossil fuel or organic contaminants within the waste may also contain gas-forming salt, Cl, as well as calcium salts from acid-gas scrubbing systems and absorbed volatile metals. These gases must be collected and sent to an offgas treatment system consisting of wet scrubbers and filters. An induced draft blower system connected at the tail end of the offgas treatment system keeps the operating pressure of the vitrification process slightly negative with respect to atmospheric pressure. Therefore, any gaseous leaks during processing will be inward, providing greater assurance of protection to workers and the public during operation.

There are two types of electrical vitrification processes – refractory-lined melters and in situ vitrification (ISV).

Refractory-Lined Melters. In refractory-lined melters, the molten glass cavity, which processes the waste, is contained within a high temperature ceramic refractory. Waste constituents, which can be slurries, wet or dry solids, or combustible material, are first mixed with glass formers and then conveyed onto the surface of the molten glass. Offgases due to evaporation of water, chemical decomposition, particulate entrainment, and combustion are scrubbed and/or filtered before being released to the atmosphere. The electrodes, which are commonly flat plates placed at either end of the melting cavity, or as with electric arc furnaces large diameter rods, keep the glass pool molten as feed material is introduced to the surface of the melt. The electrodes are usually composed of a nickel/chromium alloy, but can also consist of graphite or molybdenum rods. Glass formed during the vitrification process can be withdrawn continuously or through an overflow weir into the receiving canisters, drums, or glass quenching systems.

When combustible materials are present, additional oxygen or air is introduced into the cavity of the open pool. The feed rate and melter design can be adjusted to control the residence time for efficient destruction of the organic contaminants.

Different electric melter designs, primarily the horizontal melter and the vertical melter, are available. A horizontal glass melter is a relatively long and shallow chamber. Waste is introduced with the glass formers at one end of the chamber and the offgases and glass product are removed at the other. This type of melter is designed to provide long residence time for destruction of combustible gaseous effluents. The vertical glass melter has a pooled molten glass chamber that is roughly equal in length and width. The waste is introduced near the center of the melt.

When waste constituents are vitrified, there is generally a volume reduction due to the consolidation and densification of the material. For contaminated materials that do not require a great deal of additional glass formers, the volume reduction is generally in the range of 10 to 30%. In the case of dry combuster ash residue from municipal wastes, the volume reduction approaches 80%.

Because heat is distributed within the glass melt, electric glass melters provide the advantage of being able to be scaled to support the design process rate. When fed liquids and slurries, vertical glass melters have been demonstrated at rates of 30 to 100 L/hr/m^2 (0.7 to 2.5 gal/hr/ft^2) of top surface area (Freeman 1988). The solid material processing rate is 80 to 120 kg/hr/m^2 (16 to 24 lb/hr/ft^2) (US EPA 1992b). Therefore, a vertical melter capable of processing 220 kg/hr (480 lb/hr) would have to have internal dimensions roughly 1.4 m (4.5 ft) on a side. Similarly, a horizontal melter 1.2 m (4 ft) wide by 7 m (23 ft) long is sufficient to process dry solids at a rate of 220 kg/hr (480 lb/hr) (Freeman 1988). A 2.7 m (8.9 ft) by 1.2 m (3.9 ft) deep, water-cooled electric melter can accommodate a processing rate of 3,600 kg/hr (7,940 lb/hr) for dry municipal waste incinerator ash.

Operational costs for an electric heated glass melter have been estimated by Chapman (1991) at $0.06/kg ($53/ton) of dry municipal incinerator ash, a nonhazardous waste. This estimate is based on a 55 tonne/day (50 ton/day) melter. This process is quite inexpensive because of reduced energy requirements; residual carbon present in the ash acts as a supplemental heat source upon combustion. The estimate takes into account the advantage of other ongoing operations also, such as offgas treatment associated with the incinerator. Operational costs for electric, refractory-lined melters are highly dependent on the kind of waste being vitrified. For example, the Bureau of Mines estimates costs at $130 to $225/dry tonne ($115 to $205/dry ton) of nonradioactive, solid waste material. Freeman (1988) estimated the cost of treating low-level nuclear waste in the range of $0.63 to $1.92/kg ($0.29 to $0.87/lb) of wastes, and Ross and Kindle (1992) estimated the costs for mixed radioactive waste treatment to be $1,600/m^3 ($1,200/yd^3). For chemical hazardous wastes, these costs would probably be drastically reduced because there would be no need for remote operation features as there is when vitrifying nuclear waste material. Koegler et al. (1989) estimated the total cost of vitrification of hazardous waste refractory-lined melters to be $770/tonne ($700/ton), including offgas treatment by wet scrubbing and filtration.

In Situ Vitrification (ISV). In processes employing refractory-lined melters, wastes are fed to a molten pool within a refractory-lined cavity; ISV, treating contaminated soils and sludges in place, eliminates the need for refractory lining. In situ vitrification is a batch process, whereas those using refractory-lined melters are continuously-fed processes. To begin the process, four graphite electrode rods are inserted vertically a few centimeters into the surface of the soil. A mixture of ground glass frit and graphite flakes is laid in an "X" pattern among the four electrodes. Voltage applied to the electrodes initiates an electrical current within the starter path that heats the surrounding soil and causes it to melt. Once molten, the soil begins to conduct electrical current and the graphite is consumed by oxidation.

The molten soil grows outward and downward until the desired vitrification depth is obtained (Buelt et al. 1987). Figure 3.2 illustrates the process.

The processing rate for ISV is generally 2.7 to 4.5 tonne/hr (3 to 5 ton/hr), and the process is capable of producing blocks up to a 910 tonne (1,000 ton) per batch setting. The blocks can measure up to 12 m (39 ft) on each side. For larger sites, multiple settings would be used, and the blocks would be fused together.

The maximum demonstrated depth of the process has been 6 m (19 ft), but with enhanced depth techniques currently under development, the projected depth is greater than 10 m (33 ft). The depth, however, can be limited by the presence of heterogeneous conditions, such as layers of rock or higher melting soil compositions, or high groundwater tables.

As with the refractory-lined melters, the projected volume reduction is about 30%; that is, the vitrified volume is 70% of the initial volume of the soil. The subsidence void that is evident on the surface after processing is simply backfilled with clean soil to restore the site to its original surface level. The volume reduction for sludges of high combustibility and/or high

Figure 3.2
In Situ Vitrification

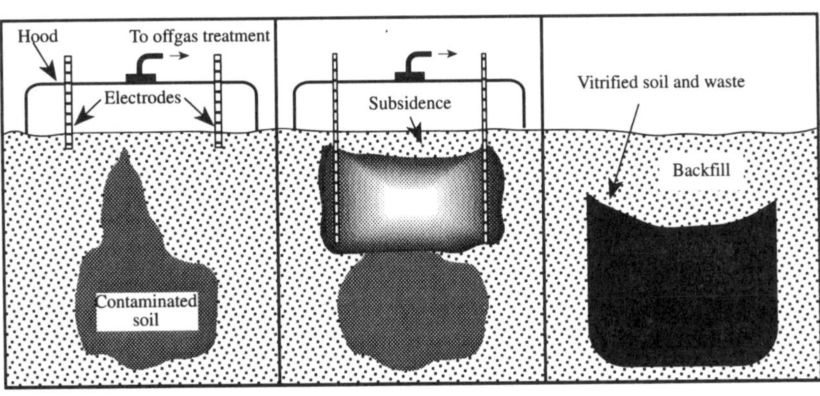

During in situ vitrification, the soil melts from the surface to the desired depth, producing a strong, high-integrity residual that can safely withstand long-term environmental exposure.

moisture content can be much greater. A volume reduction of more than 3:1 (70%) has been demonstrated with sludges containing 55% water by weight (Buelt and Freim 1986). When contaminated soils extend into a permeable aquifer, however, the recharge of moisture to the vicinity of the melt can prevent melting into the saturated zone. It is projected that permeabilities greater than 10^{-5} cm/sec will impede the progress of the melt into a saturated zone (Buelt et al. 1987).

The ISV technology is presently commercially available for use on contaminated soils of virtually all types. The estimated costs of vitrifying contaminated soils when translated into 1992 dollars, is approximately $385 to $495/tonne ($350 to $450/ton) of material, which includes offgas treatment, energy, labor, materials, and an amortized amount for capital equipment and equipment mobilization (Buelt et al. 1987).

3.4.3.2 Thermal Vitrification

Unlike electrical vitrification, thermal vitrification uses an external heat source, such as fossil fuel combustion, to melt the material. The heat is then transferred by convective and radiative heat transfer to the mixture of waste constituents, glass formers, and fluxes. Although other types of thermal vitrification exist, such as the cyclone furnace (US EPA 1991), the primary technique for thermal vitrification is the rotary kiln incinerator (US EPA 1992b). The rotary kiln is a cylindrical, refractory-lined shell that is mounted at an incline and rotated slowly to drive the materials through the incinerator. The rotation also helps improve heat transfer and facilitates mixing of the materials during the vitrification process. Waste materials and fuel are injected at the upper end of the rotary kiln. Combustible materials are oxidized within the combustion chamber. The vitrified ash is withdrawn from the lower end of the kiln. A typical rotary kiln is operated at 150 tonne/day (165 ton/day) at temperatures of about 1,200°C (2,200°F) (US EPA 1992b). Rotary kilns with production rates on this order are generally 60 to 90 m (200 to 300 ft) in length.

The cyclone furnace is emerging as a vitrification process for hazardous wastes from well-established coal-burning technology (Batdorf, Gillins, and Anderson 1992). In this process, pulverized solid waste (less than 6 mm (0.24 in.)), such as soil, is introduced with fossil fuel in a cyclic chamber. The molten slag that is formed is collected on the walls of the furnace and is then tapped through a slag spout. Cyclone furnace designs generally quench the molten slag in water, which must then be subsequently treated periodically. The organic components associated with the feed material, besides adding to the heating value to support the combustion process, are destroyed in the one, or as in some designs, two stages in primary and secondary chambers.

Information on operating costs of thermal vitrifiers is not documented in available sources.

3.4.3.3 Plasma Vitrification

Plasma systems can be used to incinerate and vitrify contaminated materials by passing them through a plasma arc or torch that creates extremely high temperatures for destruction of liquid and solid waste. The high temperatures in the plasma are created by an electrical discharge through a gas. The resulting plasma destroys the waste through radiant heat transfer. As with other vitrification processes, the offgas is passed through a treatment system that can consist of a caustic scrubber to remove acid gases and ash components, followed by filters. Typical plasma torches and arc melters have power requirements of about 600 kw, although larger systems with high-production rates are emerging rapidly (Donaldson, Carpenedo, and Anderson 1992). This translates to production rates ranging from 4 to 19 kg/min (9 to 42 lb/min) or up to approximately 0.45 tonne/hr (0.5 ton/hr) (Pacific Northwest Laboratory 1991).

Basically, two types of plasma vitrification processes are available, plasma torch, in which waste is passed through two water-cooled, generally copper electrodes, and plasma arc, in which an electrical arc is passed in the near vicinity or in contact with a molten slag. One example for the plasma torch process is the plasma centrifugal furnace (US EPA 1991). Contaminated soils enter a sealed, rotating furnace. The centrifugal forces cause the soils to cling to the sides of the furnace, exposing them to the plasma arc. Organic contaminants are destroyed at 1,300°C (2,400°F) and the molten soil at up to 1,600°C (2,900°F). The soil is periodically removed (tapped) and allowed to cool in the slag chamber. One example of the plasma arc process is the graphite electrode DC arc furnace (Surma et al. 1993). This furnace shown in figure 3.3 (on page 3.16), is composed of an insulated graphite pot, which contains the molten slag, and a hollow graphite electrode above the surface of the molten slag. Offgases are drawn through the hollow electrode, thus exposing them to maximum temperatures to maxi-

Figure 3.3
Pilot-Scale, Graphite-Electrode DC Arc Furnace

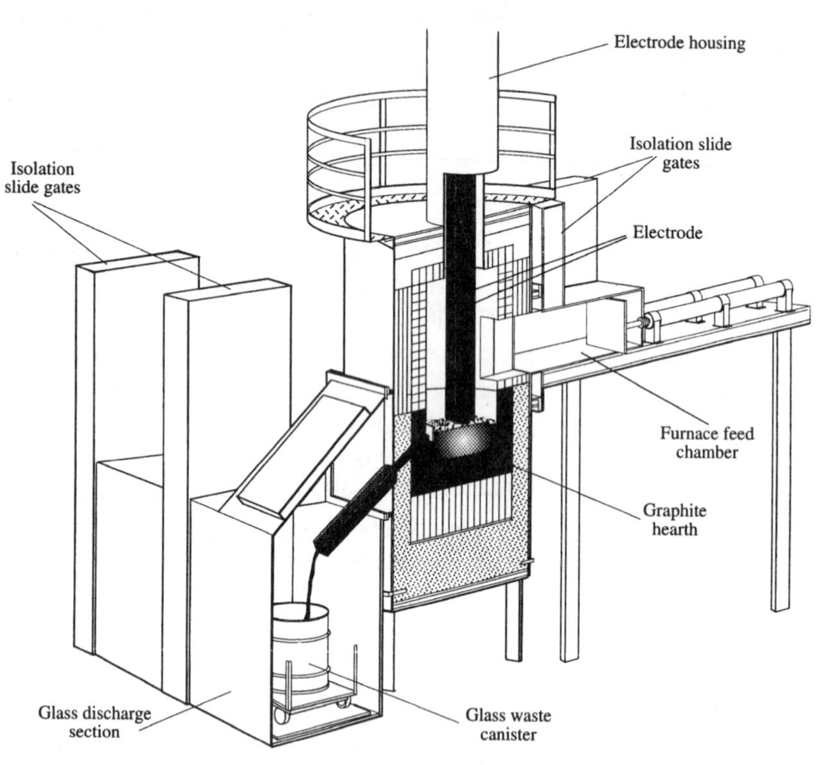

mize organic destruction efficiency. The electrical arc can be established in the nontransferred mode, in which the electrical arc is created just above the molten material being treated, or the transferred mode, which establishes the arc between the electrode and the molten slag itself. The former operating mode assists in startup of the process, while the latter is used to improve operating efficiency.

In comparison with thermal vitrification systems that require fossil fuels for heat generation, the plasma system accomplishes heat transfer much faster because of the low gas-flow requirements and extremely high temperatures of up to 5,500°C (10,000°F) (US EPA 1992b). The plasma process allows vitrification to take place in a relatively small piece of equipment. For example, a transportable pilot-scale system designed for destruction of liquid wastes has been housed in a 14 m (45 ft) trailer.

No operational cost information is available for plasma vitrification systems in any of the sources cited in the List of References, Appendix A.

3.5 Modified Sulfur Cement Process

3.5.1 Description

Modified sulfur cement was developed by the United States Bureau of Mines in 1972 as a means of utilizing waste sulfur from flue gas and petroleum distillation processes. Previous attempts to use elemental sulfur as a construction material in the chemical industry (Raymont 1978) failed because of internal stresses set up by changes in crystalline structure during cooling. By reacting elemental sulfur with hydrocarbon polymers, the Bureau of Mines developed a product that successfully suppresses the solid phase transformation, and thus dramatically improves stability of the material. A Bureau of Mines formulation that is licensed commercially (Martin Resources, Inc., Odessa, Texas) was used for this work. The formulation contains a total of 5% by weight modifiers consisting of equal amounts dicyclopentadiene (DCPD) and cyclopentadiene (CPD) that react with the sulfur to form long chain polymers (Sullivan and McBee 1976). It has a melting point of 119°C (246°F) and a viscosity of approximately 25 centipoise (cp) at 135°C (275°F).

Modified sulfur cement is a thermoplastic material that can be easily melted, combined with waste components in a homogeneous mixture, and cooled to form a solid, monolithic waste form. Compared with hydraulic cements, sulfur cement has several advantages. For example, no chemical reactions are required for solidification, eliminating the possibility that elements in the waste can interfere with setting and thereby limit the range of waste materials that can be successfully encapsulated. Sulfur concrete compressive and tensile strengths twice those of comparable portland concretes have been achieved, and full strength is attained in several hours rather than weeks (Sulfur Institute 1979). Sulfur concretes are resistant to attack by most acids and salts, e.g., sulfates that can severely degrade hydraulic cement, and have little or no effect on the integrity of sulfur cement (McBee, Sullivan, and Jong 1985).

3.5.2 Operational Considerations

The first application of modified sulfur cement to the solidification of radioactive and mixed wastes was performed at Brookhaven National Laboratory (Colombo, Kalb, and Fuhrmann 1983). Processing of modified sulfur cement and wastes to form a solid, monolithic form requires (1) heating of the components until the binder has completely melted, (2) mixing the constituents into a homogeneous molten slurry, and (3) pouring the mixture into suitable containers for storage and disposal. These relatively simple processing requirements can be met by a number of different techniques, including double planetary mixers, in-drum mixers, pug mills, and in general, the same type of equipment used to make hot asphalt mixes.

3.5.3 Health and Safety Considerations

When sulfur cement materials are produced in the recommended mixing temperature range of 127° to 149°C (260° to 300°F), gaseous emissions of sulfur dioxide and hydrogen sulfide will not exceed the allowable threshold limit values and sulfur vapor emissions will be minimized. The threshold values established for sulfur dioxide are 5 ppm for a short-term exposure and 2 ppm (time-weighed average concentration) for an 8 hr exposure. The corresponding values for hydrogen sulfide are 15 and 10 ppm (McBee, Sullivan, and Fike 1985).

3.6 *Polyethylene Extrusion Process*

3.6.1 Description

A polyethylene extrusion process for treatment of radioactive, chemical hazardous, and mixed wastes has been developed at Brookhaven National Laboratory. The extrusion process for the encapsulation of wastes in polyethylene involves the heating, mixing, and extruding of materials in one basic operation. To more clearly understand this process, it may be broken down into the following steps (Kalb and Colombo 1984):

- The polyethylene binder and predried waste materials are transferred from either a single hopper or individual hoppers in which they are stored to the extruder feed throat. Metering of waste-to-binder ratios is accomplished at this step;
- The mixture is conveyed through a heated cylinder by the motion of the rotating screw. The initial portion of the cylinder is controlled at a temperature below the polyethylene melting point (120°C (250°F)). This serves to gradually preheat the materials, but, at the same time, assure proper transport of the mixture;
- As the waste-binder mixture moves forward past the initial preheating zone, it is masticated under pressure due to the compressive effects of a gradual reduction in the channel area between the screw and cylinder. Screw rotation also assists in the mixing of the materials to a homogenous state;
- The gradual transfer of thermal energy by the combined effects of the barrel heaters and frictional heat serves to melt the mixture. The fractional heat input is difficult to control and must be compensated for by the regulation of the resistance band heaters. In some cases, it is necessary to remove excessive heat by use of external blowers or coolers; and
- The melted mixture is forced through an output die into a mold where it is allowed to cool and solidify.

An extruder consists of four basic components, as depicted in the simplified schematic diagram in figure 3.4 (on page 3.20). These are (1) a feed hopper, (2) a rotating auger-like screw, (3) a heated cylinder in which the screw rotates, and (4) an output die assembly to shape the final product.

3.6.1.1 Polyethylene

Polyethylene is a thermoplastic, organic polymer of paracrystalline structure formed through the polymerization of ethylene gas. Thermoplastic polymers consist of branched or linear polymer chains that normally are not cross-linked. At elevated temperature, thermoplastic polymers change from a hard material to a rubbery, flowable liquid. On cooling, the polymers revert to their original form.

Polyethylene has been commercially produced in the United States for over 45 years and is the most widely used of all polymers. Current production capacity for the ten domestic producers of low density polyethylene (LDPE) is about 3 x 10^8 kg/yr (7 x 10^8 lb/yr). The market price for LDPE is approximately \$0.84/kg (\$0.38/lb) (Theis 1989).

Mechanical and physical properties of LDPE are listed in table 3.2 (on page 3.21), along with the test methods used to determine them. These data provide an overview of the strength and durability of this material and background information concerning potential performance for encapsulation of various kinds of waste.

Figure 3.4
Sectional View of a Simplified Screw Extruder

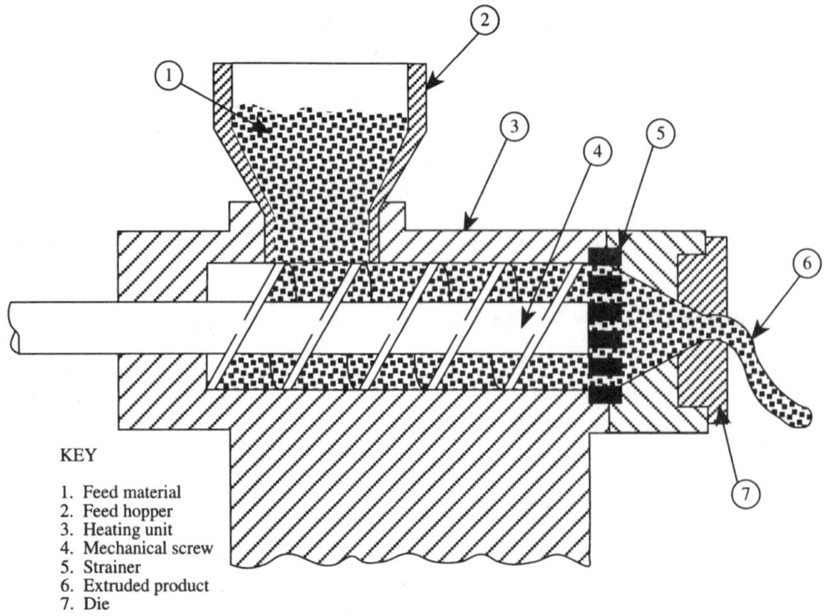

The sketch depicts flow of material from the hopper to the output die, where it is extruded in a molten state.

Table 3.2
Properties of Low Density Polyethylene[a]

Property	ASTM Test Method	Results
Brittleness temperature, °C	D 746	-80 to -55
Burning rate	D 635	very slow
Crystalline melting point, °C		108-126
Deflection temperature, at 66 psi, °C	D 648	65-80
Density, g/cm^2	D 792	0.910-0.925
Elongation, %	D 638	100-650
Extrusion process temperature, °C		121-232
Flexural modulus, at 23°C, 10^3 psi	D 790	35-48
Hardness, Shore	D 2240	Shore D 44-50
Heat capacity, cal/°C/g		0.55
Heat resistance, continuous, °C		82-100
Impact strength, ft lb/in of notch	D 256	>16 (No break)
Melt index, g/10 min	D 1238	0.2-55
Mold (linear)shrinkage, in/in		0.015-0.050
Molecular weight, weight average		10^4 - 10^5
Organic solvents, effect of		resistant below 60°C
Tensile impact strength, ft lb/in^2	D 1822	180
Tensile modulus, 10^3 psi	D 638	25-41
Tensile strength, at break, psi	D 638	1,200-4,550
Tensile yield strength, psi	D 638	1,300-2,100
Thermal conductivity, 10^{-4} cal cm/sec cm^2 °C	C 177	8
Thermal expansion, 10^{-6} in/in/°C	D 696	100-220
Water absorption, (1/8 in specimen), 24 hr., %	D 790	<0.01

(a) Adapted from Theis 1989 and Bikales 1967

Polyethylene's resistance to aggressive chemicals is a primary reason for its widespread use in many diverse applications. The key to ensuring long-term stability of waste forms is the ability of the binder to withstand the chemical environment, internally, of the waste materials, and externally, of the disposal site. Protection from a broad range of chemical reagents is essential because it is difficult to accurately predict the chemical composition of leachates resulting from varied waste forms in disposal.

At ambient temperatures, polyethylene is insoluble in virtually all organic solvents and is resistant to many acids and alkaline solutions (Raff and Allison 1956). Polyethylene is resistant to any concentration of hydrochloric, hydrofluoric, phosphoric, and formic acids, ammonia, potassium hydroxide, sodium hydroxide, potassium permanganate, and hydrogen peroxide. In dilute concentrations (up to 50% by weight), polyethylene is resistant to sulfuric and nitric acids. Exposure to some aliphatic, aromatic, and chlorinated hydrocarbons can cause swelling, but will not permanently change the mechanical properties of polyethylene (e.g., strength), because the original properties reappear upon evaporation of the swelling medium. In general, LDPE is relatively unaffected by polar solvents, including alcohols, phenols, esters, and ketones. Table 3.3 (on page 3.23) provides an overview of polyethylene's compatibility with many chemicals, after three months' contact.

3.6.2 Status of Development

3.6.2.1 Bench-Scale Development

The feasibility of encapsulating various types of wastes in polyethylene has been demonstrated for sodium nitrate salts, incinerator fly ash, ion-exchange resins, and evaporator concentrates from nuclear facilities (Franz and Colombo 1985).

Formulation development work was conducted using a bench-scale 32 mm (1.25 in.) polyethylene extruder, with a maximum output capacity of about 16 kg/hr (35 lb/hr), shown in figure 3.5 (on page 3.24). Envelopes of potential and optimal operating ranges were defined for critical process parameters, such as, temperature, pressure, feed rate, extrusion rate, and solidification kinetics. The effect of other parameters, including polyethylene flow properties (i.e. melt index), waste-binder mixing techniques, waste pretreatment requirements, and power requirements were also investigated. Some of these parameters are reviewed briefly here in order to provide background for the discussion of scaleup feasibility in Section 3.6.2.2 (Kalb and Colombo 1984).

Temperature. Temperature is an important parameter because it affects processibility and product quality. For waste encapsulation, minimal temperatures are preferred in order to reduce potential volatilization of contaminants. Typical process temperatures for waste encapsulation ranged between 125° and 150°C (257° to 300°F).

Pressure. Pressure is a function of many factors, including the kind of polyethylene used, waste characteristics, and extrusion parameters, such as

Table 3.3
Effect of Various Chemicals on Polyethylene after 3-Month Contact [(a)]

Reagent	On Removal from Reagent: Change in Wt, %	On Removal from Reagent: Appearance	After 24 Hr. Conditioning at Room Temp.: Tensile Strength, psi	After 24 Hr. Conditioning at Room Temp.: Elongation, %	Resistance Rating [(b)]
Inorganic Acids and Bases					
H_2SO_4, conc.	+ 0.13	No change	1458	462	E
H_2SO_4, 10%	+ 0.04	No change	1370	483	E
HCl, conc.	+ 0.13	No change	1406	258	G
HCl, 10%	+ 0.20	No change	1442	336	E
HNO_3, conc.	+ 3.02	No change	1093	71	F
HNO_3, 10%	+ 0.22	No change	1387	325	E
NaOH, 50%	+ 0.13	No change	1432	313	E
NH_4OH,conc.	+ 0.31	No change	1378	371	E
Oxygenated Organic Compounds					
Ethanol (denatured)	- 0.02	No change	1550	421	E
Acetone	+ 0.03	No change	1363	379	E
Ethyl acetate	+ 2.76	No change	1295	325	E
Dioxane	+ 0.38	No change	1368	382	E
Butyraldehyde	+ 3.06	No change	1245	417	E
Linseed oil	+ 0.88	No change	1410	483	E
Triethanolamine	+ 0.08	No change	1408	379	E
Camphor Oil	+17.42	Swollen	1375	483	G
Hydrocarbons					
Ethyl gasoline	+11.75	Swollen	1430	508	G
Benzene	- 0.86	No change	1464	429	E
Xylene	- 0.70	No change	1623	479	E
Lubricating oil	+ 7.54	Swollen slightly	954	167	F
Carbon tetrachloride	+22.35	Swollen	1560	475	E
Ethylene dichloride	+ 0.80	No change	1526	300	E
Tricholorobenzene	+26.33	Swollen	1450	442	G
Aqueous Solutions of Salts					
Sodium bisulfite, 10%	+ 0.17	No change	1310	483	E
Calcium chloride, 15%	+ 0.70	No change	1380	483	E
Calcium hypochlorite (bleaching soln.)	+ 0.06	No change	1330	375	E
Duponol ME (fatty alcohol sulfate), 10%	+ 0.04	No change	1225	463	G
Ferric sulfate, 15%	+ 0.02	No change	1307	467	E

(a) Adapted from Raff and Allison 1956

(b) E = excellent; G = good; F = fair

Figure 3.5
Photograph of Laboratory-Scale Extruder with Separate Dynamic Feeders for Waste and Binder

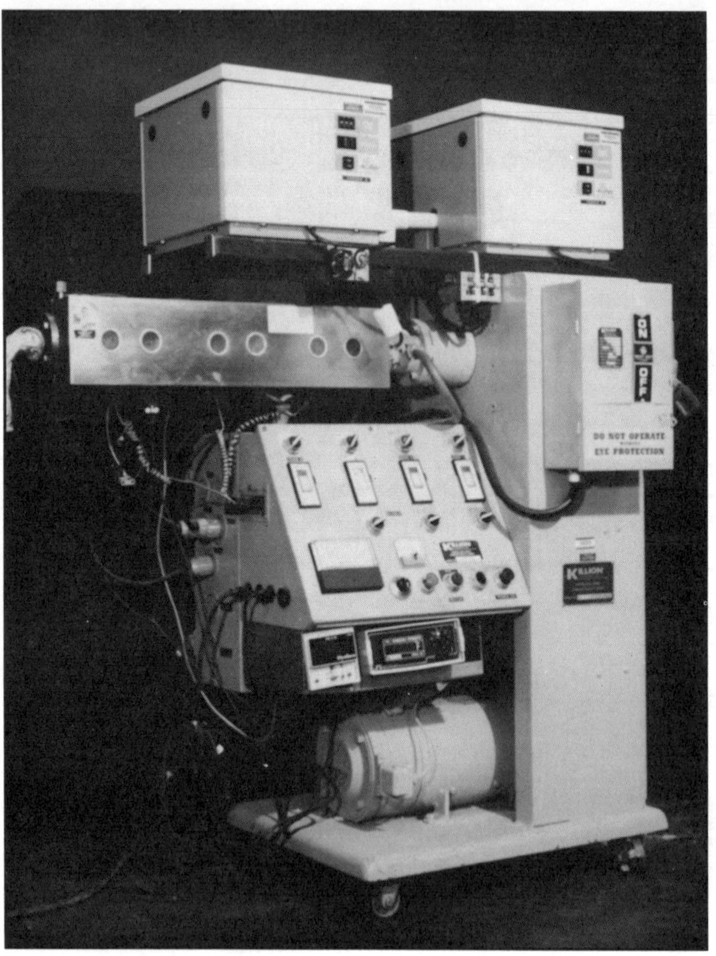

Franz, Heizer, and Colombo 1987

screw design, temperature, and rate. Moderate pressures in the range of 1 to 20 MPa (up to several thousand psi) are desirable, because they enhance mixing and the delivery of mixture to the mold.

Process Rates. Feed and extrusion rates can vary depending on waste loading and characteristics; they must be coordinated to avoid jamming or starving. The laboratory-scale extruder was operated with an output rate between 1 kg/hr (2.2 lb/hr) and 9 kg/hr (20 lb/hr).

Melt Index. Melt index, shown among other properties of LDPE in table 3.2 (on page 3.19), is the measure of a material's capacity for flowing at 190°C (374°F). For optimal processing of laboratory-scale waste forms, a melt index of 55 g/10 min (1.9 oz/10 min) was selected.

Feed Method. When using a stock (static hopper) feed system, density and particle size differences led to segregation of waste and binder, creating a heterogeneous mixture. Separate dynamic feeders for waste and binders were used to overcome these difficulties and provide a means for precisely controlling waste-binder ratios.

Pretreatment. For optimal extrusion, waste should be dry; granular particles provide the best flow characteristics.

3.6.2.2 Scaleup Feasibility

When bench-scale formulation and testing for the polyethylene encapsulation of nitrate salt was complete, planning was initiated to demonstrate this system using production-scale equipment in conjunction with Rocky Flats Plant (Kalb, Heiser, and Colombo 1992). Estimates of necessary production capacity were made based on data from the Rocky Flats Plant. Sizing of extruder equipment varies over a wide range, from bench-scale equipment with a screw diameter of 19 mm to 32 mm (0.75 to 1.25 in.) to very large machines with screw diameters up to 152 mm (6 in.) or more. Typically, laboratory-scale extruders can process around 9 to 14 kg/hr (20 to 30 lb/hr), while production-scale machines can process hundreds or thousands of kg/hr. For the technology demonstration, a 114 mm (4.5 in.) extruder, with output capacities in the range of 900 kg/hr (2,000 lb/hr) was selected. Data on the process parameters generated during bench-scale investigations were reviewed to develop a set of required design specifications. A survey of potential vendors was then conducted. Equipment designs were examined in order to assure that processing and monitoring requirements could be met using conventional, off-the-shelf equipment.

The production-scale feasibility test was conducted using a 114 mm (4.5 in.) extruder at laboratory facilities provided by Davis-Standard (Pawcatuck, Conn.), a manufacturer of extruder equipment whose facility is designed for feasibility testing. It is equipped with state-of-the-art monitoring and control systems for data collection and process control, as well as a wide selection of extruder and screw designs to accommodate diverse user needs. Personnel from Brookhaven National Laboratory (BNL) and Rocky Flats Plant attended the demonstration; the staff at Davis-Standard provided technical assistance.

Two Accu-Rate (Whitewater, Wis.) Model 610 dry material feeders were purchased and calibrated prior to the feasibility test in order to maintain an accurate waste loading of sodium nitrate salt. A slightly more conservative waste loading of 60% by weight nitrate salt was selected for the scaleup feasibility test (compared to the maximum loading of 70% by weight) because of the many variables under consideration. (Optimization of waste loading under full-scale conditions will be performed as part of the planned Technology Demonstration.) These feeders are larger versions of the feeders used in bench-scale studies. They were installed above the extruder feed throat, in place of a standard static feed hopper. Figure 3.6 (on page 3.27) is a simplified process flow diagram of the full-scale polyethylene encapsulation system.

Initial trials resulted in excessive foaming due to air entrainment, caused by the physical properties of the simulated granular salt waste and length of the extruder barrel. This problem was remedied quickly by removing the extruder screw and replacing it with one designed to vent gases midstream. Process settings similar to those developed at BNL, shown in table 3.4 (on page 3.27), were successfully duplicated. Maximum output rates were not attempted, but an output peak of 1,577 kg/hr (3,477 lb/hr) was attained at 65 rev/min and steady-state rates in excess of 454 kg/hr (1,000 lb/hr) were easily achieved. A 114 L (30 gal) drum of encapsulated sodium nitrate was filled in about 25 minutes. Only minimal shrinkage was observed upon cooling, indicating a lack of significant voids in the waste form.

Conclusions resulting from the production-scale feasibility test for the polyethylene encapsulation of nitrate salt wastes may be summarized, as follows:

- Polyethylene encapsulation of at least 60% by weight nitrate salt wastes can be accomplished successfully, using a production-

Figure 3.6

Polyethylene Encapsulation System Process Flow Diagram

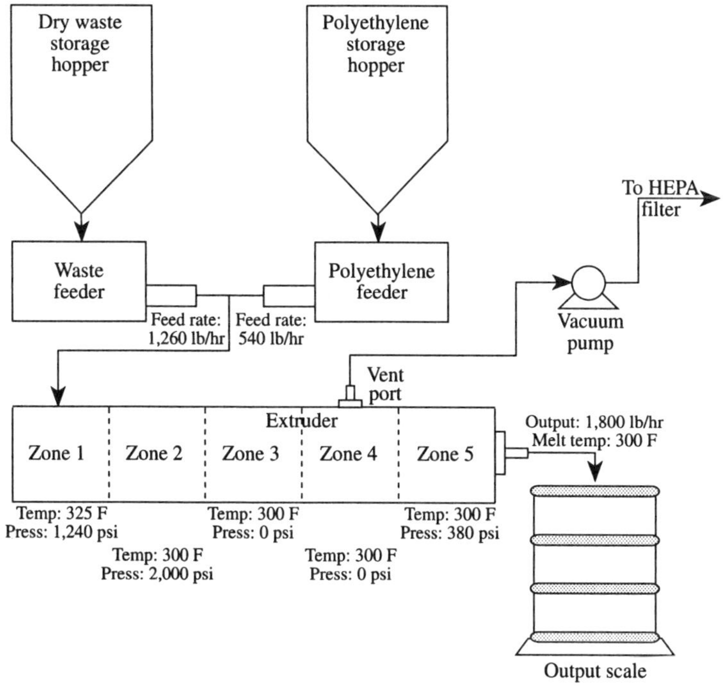

Table 3.4

Parameter Settings and Test Data for Polyethylene Encapsulation Production-Scale Feasibility

Melt Temperature, °C (°F)	149 (300)
Melt Pressure, MPa (psi)	2.6 (380)
Max. Screw Speed, RPM	65
Steady-state Output, kg/hr (lb/hr)	>454 (1,000)
Max. Output, kg/hr (lb/hr)	1,577 (3,477)
Horsepower at Max Output	354

scale 114 mm (4.5 in.) extruder at steady-state rates of at least 454 kg/hr (1,000 lb/hr);

- Comparison of bench- and production-scale process data confirms that process scaleup is feasible and that information generated during research and development phases of this project can be applied during the technology demonstration phase;
- Quality assurance testing of the 114 L (30 gal) waste form produced during the full-scale feasibility test demonstrates that a homogenous waste form with excellent properties can be produced using off-the-shelf production equipment; and
- Close agreement of results from waste form performance tests of specimens produced using bench- and full-scale systems, indicates the validity and importance of bench-scale research and development, prior to scaleup and demonstration of new technologies.

3.7 *Inorganic, Cementitious Technologies of the Siliceous Category*

3.7.1 Description

Following are characteristics of inorganic, cementitious stabilization/ solidification systems:

- relatively low cost;
- good, long-term stability, both physical and chemical;
- documented use on a variety of industrial wastes over a period of at least ten years;
- widespread availability of the chemical ingredients;
- nontoxicity of the chemical ingredients;
- ease of use in processing (processes normally operate at ambient temperature and pressure and without unique or very special equipment);

- wide range of volume increase;
- inertness to ultraviolet radiation;
- high resistance to biodegradation;
- low water solubility;
- relatively low water permeability; and
- good mechanical and structural characteristics.

3.7.2 Scientific Basis

Setting and curing reactions vary among processes. Most of the commercial, cementitious inorganic stabilization/solidification systems, however, solidify in highly similar reactions, which have been thoroughly studied in connection with portland cement technology used in concrete making. While the pozzolanic reactions of the processes using fly ash and kiln dusts are not identical to those of portland cement, the general reactions are alike. One reason for this is presented in an interesting way by Cote (1986). The compositions of most of the primary reagents used in inorganic stabilization/solidification systems were plotted on a ternary diagram using the three oxide combinations, SiO_2, CaO + MgO, and Al_2O_3 + Fe_2O_3. All of these reagents have the same active ingredients as far as solidification reactions are concerned; this is illustrated in figure 3.7 (on page 3.30). The combinations of these five oxides express the essential composition of any of these materials, even though, in many cases, the actual compounds are not simple oxides, but more complex silicates and aluminates. It is interesting that all of the reagents, with exception of power-plant fly ash, have their origin in natural limestone and clay formations, and all are inexpensive compared to industrial "chemicals."[1]

3.7.3 Operational Considerations

The cementitious reactions that occur in these processes require the pH to be above 10. This level is achieved through the dissolution of free lime from the solid, and continues throughout the setting and curing stages of the mixture. A "false set" can occur when enough free lime is present initially

1. Reprinted by permission of Chapman & Hall from "Chemical Fixation and Solidification of Hazardous Wastes" by Jesse R. Conner. Copyright 1990 by Chapman & Hall.

Figure 3.7
Ternary Composition Diagram for Cementitious Systems

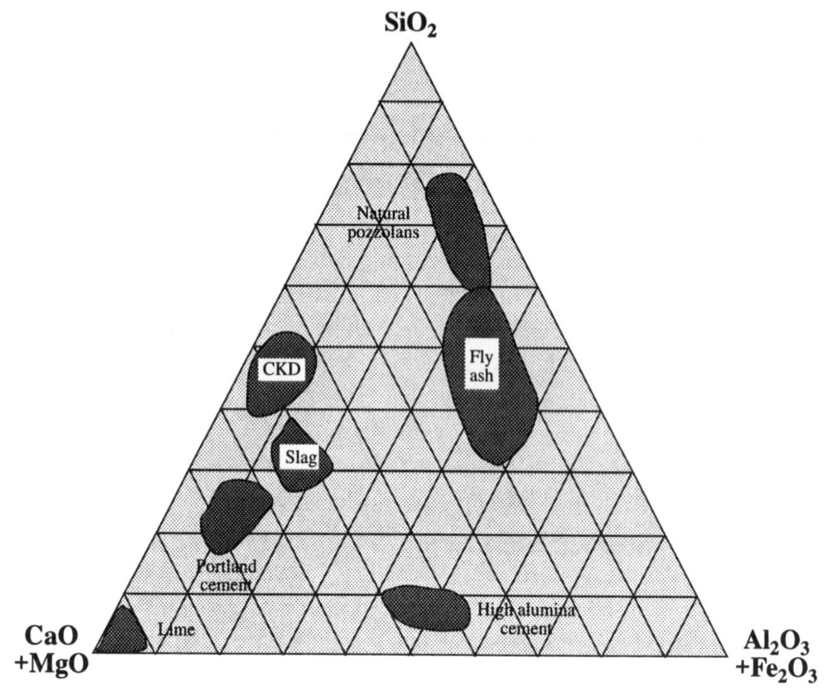

Adapted from Conner 1990

to start the reactions, but availability of free lime decreases as the process proceeds and the reactions stop or slow down. Therefore, any reaction that competes successfully for the calcium ion may inhibit setting. The cementitious or pozzolanic reactions also require sufficient free water if they are to run to completion. Water is used up in the hydration reaction of these chemical systems. The water content, as measured by total solids determination, is not necessarily all available for reaction.[2]

2. Reprinted by permission of Chapman & Hall from "Chemical Fixation and Solidification of Hazardous Wastes" by Jesse R. Conner. Copyright 1990 by Chapman & Hall.

The inorganic processes include those that use bulking agents, such as Class F fly ash, and those that do not. A bulking agent, in this context, is an additive that primarily adds to the total solids and viscosity of the waste, thus preventing settling out of the suspended waste components before solidification can occur; it may also help produce a solid with better physical properties. Examples of these two groups of processes are: (1) cement-based or cement/soluble silicate systems (no bulking agents) and (2) cement/fly ash, lime/fly ash, cement/clay, or lime/clay (systems with bulking agents). There are two types of bulking agents: those that are essentially inert in the system and act as described above and those that also have reactive capacity or exhibit pozzolanic activity. A pozzolan is defined as a material that does not exhibit cementing ability when used by itself, but when used in combination with other materials, such as, portland cement and lime, will interact with these agents resulting in a cementitious reaction.

The most noticeable difference resulting from the use of systems with and without bulking agents is that systems with bulking agents often yield lower chemical costs because some of the more expensive cementing materials are replaced by less expensive waste products such as fly ash. Systems without bulking agents sometimes yield lower overall costs because of the lower weight and volume increase associated with them.

3.7.4 Status of Development

Conventional inorganic chemical stabilization/solidification processes that have been used commercially are listed below. The most important systems today are marked with an asterisk:

- portland cement-based (major ingredient is cement)*;
- portland cement/lime;
- portland cement/clay;
- portland cement/fly ash*;
- portland cement/soluble silicate*;
- lime/fly ash*;
- cement or lime kiln dust*; and
- slag.

Many of the permutations and combinations of these systems, including those varying as to the kinds of additives used, are patented or are covered by patent applications, but information about most of the generic systems is believed to be in the public domain. All of these processes have been used commercially for solidification of water-based waste liquids, sludges, filter cakes, and contaminated soils. A large body of technical information about them concerning leachability, physical properties, and general stability is available.

In terms of volume of waste treated, the lime-fly ash process has probably been the most used in the U.S., although it has been very narrowly applied, primarily to flue-gas desulfurization sludges. For other kinds of wastes and industrial sludges, the kiln dust and portland cement-based processes are the most widely used at the present. In the United States, one of the most flexible techniques used is the portland cement-sodium silicate process. This process has been applied to a variety of wastes and a good technical base, including leaching test information, is available. Another technique, the portland cement-fly ash process, has been used in Europe and Canada, but has not been applied extensively in the United States. This process works well with certain kinds of waste; like the lime-fly ash process, it involves large additions of the solidifying agents and, therefore, results in large volume increases.

3.7.5 Technology Variations

Innovative systems at the present include those using the reagents discussed in Subsections 3.7.5.1 through 3.7.5.4, below, usually in combination with conventional cementitious systems, but sometimes alone.

3.7.5.1 Soluble Silicate Processes

Sodium silicates have been used for more than a century in the production of commercial products, such as, special cements, coatings, molded articles, and catalysts. In these mixtures, the soluble silicate is mixed with cement, lime, slag, or other sources of multivalent metal ions that promote the gelation and precipitation of silicates. Katsanis, Krumrine, and Falcone (1982) have shown that the solubility of systems containing multivalent ions in combination with soluble silicates differs significantly from precipitation of solutions composed of metal salts or silica. The presence of calcium or magnesium ions, even at low concentrations, can reduce the solubility of silica by several orders of magnitude. These results suggest that soluble silicates reduce the leachability of toxic metal ions by formation of low-solubility metal oxide/silicates and by encapsulation of metal ions in a silicate- or metal silicate-gel matrix. This characteristic is one basis for the use of soluble silicates in stabilization/solidification systems.[3]

Soluble silicates have been used both as accelerators and as anti-inhibitors for concrete and have the same function in a portland cement-based stabilization/solidification system. In the precipitant-type retarders, such as heavy metals, soluble silicate probably works by removing the metal from solution before it can precipitate on the cement grains. With retarders that operate by coating the grains, soluble silicate may function as a surfactant, emulsifying oils and flocculating fine particulates so that they remain suspended in the water phase. In any event, soluble silicate has been useful in many instances for this purpose, rather than as a fixant or gellant.

The reactions of polyvalent metal salts in solution with soluble silicates have been studied extensively over many years (Vail 1952; Iler 1979). Nevertheless, the "insoluble" precipitates that result from such interactions are not usually well characterized, especially in the complex systems representative of most wastes. Falcone, Spencer, and Katsanis (1983) explained the apparent negative effects of soluble silicates on metal leachability experienced by some other investigators. When an excess of soluble silicate over available metal ion is present in such a system, the silica species with adsorbed metal ion may remain suspended, not be filtered in the leaching protocol, and give the appearance of increased leachability. Also, large excesses of soluble silicate can so decrease the set time that it is impossible to obtain a homogeneous mass, and the resulting pore size is much larger. This counters the reduction in permeability, which is the real benefit of soluble silicate in systems where the metal has already been precipitated as another species. Davis et al. (1986) found that "treatment of a waste with sodium silicate reduced the degree to which leaching acid can penetrate the waste matrix by a factor of five to seven." The sum result of these effects is demonstrated graphically in figure 3.8 (on page 3.34). In this system, a silicate content range of 6% to 10% gives minimum leachability. Interestingly, this is precisely the range which has long been used empirically by workers in this stabilization/solidification specialty.

---

3. Reprinted by permission of Chapman & Hall from "Chemical Fixation and Solidification of Hazardous Wastes" by Jesse R. Conner. Copyright 1990 by Chapman & Hall.

Figure 3.8
Leachability of Metals as a Function of Soluble Silicate Content

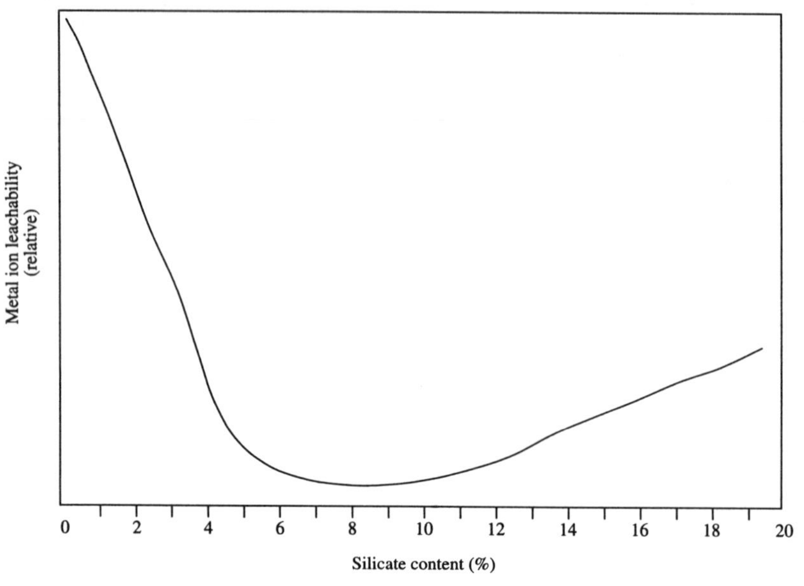

Davis et al. 1986

The current market price of sodium silicate solution, 42° Be (specific gravity), 3:2 SiO_2:Na_2O ratio, is about $180/tonne ($165/ton), FOB works in truck/car load lots. Other ratios are generally more expensive. Potassium silicate solution costs about $510/tonne ($460/ton).

Conventional soluble silicate processes or vendors are listed below:

- Chemfix Technologies, Inc., Kenner, LA (Chemfix™);
- Fujimasu Synthetic Chemical Laboratory Company, Ltd., Japan;
- Kurita Water Industries, Ltd., Japan;
- Nippon Synthetic Chemical Industry, Japan;
- SolidTek, Inc., Atlanta, GA;
- Hitachi, Japan;

- Japan Organo Co., Japan;
- PD Pollution Control, England; and
- CGE - SARP, France.

The most confusing aspect of silicate stabilization/solidification technology is that a very large number of vendors promote their processes, either intentionally or unintentionally, as "silicate" within the framework of soluble silicate systems, when, in fact, they are using other siliceous reagents. Clays, cement, slags, and kiln dusts either are or contain silicates, but not soluble silicates. Users often believe that they are buying or recommending a soluble silicate process when they are actually using one of the latter reagents. In this monograph, only true soluble silicate processes are discussed. It is possible that some processes have been missed because the vendors consider the exact nature of their processes to be proprietary.

Soluble silicate processes that appear innovative, because of either their chemistry or the way in which they are used, are described below.

EnviroGuard/ProTek/ProFix, Houston, Texas. This company sells a number of sorbents and formulated chemicals under the name enviroGuard™ (sorbent), enviroGuard Plus™ (S/S agent), and ProFix™. All of these products are based on rice hull ash, an amorphous, biogenetic silica (Durham and Henderson 1984). Because of its sorptive and alkali-reactive nature, rice hull ash has some unusual properties. Its sorptive nature is well known (Durham and Henderson 1984), but its ability to react with alkalies to form soluble silicates is of primary interest here. Under alkaline conditions, the amorphous silica reacts slowly to produce soluble silicates, which can then react with ions of toxic metals to form low-solubility metal silicates. At the same time, the soluble silicate can react with available calcium or other polyvalent metal ions to set and harden the system in a controlled manner. The advantage of this method over that of most soluble silicate processes is that the slow, continuous generation of soluble silicate provides a reserve capacity analogous to the action of buffers in a pH-control system. As metal hydroxides and other species slowly dissolve in the alkaline environment of the waste form, they can then become re-speciated as the "silicate." The process is patented in the U.S. and is the subject of patent applications in many other countries (Conner and Reber 1992).[4]

4. Reprinted by permission of Chapman & Hall from "Chemical Fixation and Solidification of Hazardous Wastes" by Jesse R. Conner. Copyright 1990 by Chapman & Hall.

A simplified version of the basic chemistry involved in the process is as follows:

$$2NaOH + xSiO_2 \rightarrow Na_2O{:}x(SiO_2) + H_2O \quad [1]$$

$$Na_2O{:}x(SiO_2) + Ca(OH)_2 \rightarrow CaSi_xO_{2_x} + 2NaOH \quad [2]$$

In place of calcium, any polyvalent metal species may be substituted, thence the formation of immobilized metals, since these metals compete with calcium for the silicate anion. The anion originally associated with the metal will help determine the reaction rate and, also, the final pH of the solid. For example, if the hydroxyl ion is dominant, sodium hydroxide will be continuously reformed to react again with the silica from the rice hull ash, until the ash is eventually exhausted. On the other hand, if the metal is in the form of chloride or sulfate, the reaction product will be neutral and alkalinity of the system will decrease until there is no longer sufficient hydroxyl ion to react with the silica. It must be understood, however, that this is a simplistic view of a very complex system, especially since the silicates formed are not exact, stoichiometric compounds.

Lopat, Wanamassa, New Jersey. Lopat has been actively marketing a process using potassium silicate and setting agents. The process has been applied in California to reduce lead leaching from automobile shredder waste. Little independent technical information has been published, although it has been said that the process can immobilize lead, dioxins, and PCBs (Environmental Science & Technology 1986). The "innovative" aspect of this process is the use of a liquid potassium silicate solutions modified with additives, K-20 (U.S. Patent 1987), as a fixative added to the waste in place of or before conventional stabilization/solidification with cementitious materials. The additives consist of "a catalytic amount of an aqueous sodium borate solution and a fixative containing solid calcium oxide," with or without a fumed silica addition. It is not stated exactly what function these additives perform in this application.

The concept of permeating soils (soil stabilization) (Joosten 1937) and other geological structures, and even waste piles (Tyco Laboratories, Inc. 1971), with soluble silicates is not new. These old methods, however, have not been commercially applied in this way until Lopat began marketing its potassium silicate formulations for the treatment of auto shredder "fluff," the residue remaining after all recyclable materials have been removed from shredded automobiles. This residue is usually classified as hazardous according to the California Waste Extraction Test (WET), but usually passes the US EPA Toxicity Characteristics Leaching Procedure (TCLP) and, therefore, is not considered to be a hazardous waste elsewhere. The Lopat process apparently was applied successfully to this waste at low cost, since physical solidification was not required.

Lopat sells the chemicals and/or licenses the process. Therefore, a number of other organizations may offer the system under the Lopat name, as "K-20," or under another name. One of these is described by Trezek (US EPA 1990a).

Other Soluble Silicate Processes. In addition to the commercial systems offered by vendors listed above, there are many patents and other technical literature that describe the use of soluble silicate processes for soil solidification, grouting, and various compositions of matter. New stabilization/solidification companies could arise in this area at any time. Most of the commercial soluble silicate processes use sodium silicate and portland cement as the chemical system. Some consider it a portland cement system with sodium silicate as an additive, while others view it as a soluble silicate system that uses portland cement as one of a number of possible setting agents for the soluble silicate. If these processes are being looked upon as being based on soluble silicates along with a setting agent, then a variety of setting agents have been enumerated in the technical and patent literature. The known chemistry of most such systems is discussed in detail in Vail (1952). Following are some of the systems:

- glycolides;
- glyoxal;
- polyalcohol esters;
- boric - phosphoric acid condensation products;
- sequestered metal ion complexes;
- succinic acid diesters; e.g., dimethyl succinate;
- methyl acetate, methyl propionate, methyl formate mixtures;
- formaldehyde or paraformaldehyde;
- diacetin and triacetin;
- formamide and ethyl acetate;

- phosphates;
- amides;
- glycerin – glacial acetic acid reaction products;
- chlorides, sulfates and nitrates of aluminum, magnesium, and iron;
- soluble polyvalent metal compounds in general;
- silicon polyphosphate;
- potassium silicofluoride and sodium silicofluoride acids; and
- mono-, di-, and triacetate acid esters of glycerol.[5]

Several patented processes in the soluble silicate stabilization/solidification area (Conner 1985, 1986a) may become commercially significant, although they are not yet commercialized. One of these is interesting because it uses a combination of liquid and dry sodium silicates to solve the problem of too-rapid setting of soluble silicate systems. In a batch-type process, which is usually the preferred approach for small quantities of wastes, the liquid silicate system sets too rapidly. This process uses small amounts of silicate solution to thicken (but not gel to a nonflowable state) the waste so that low-solids wastes will not separate before they can set, along with a powdered sodium silicate and portland cement that react more slowly to set and harden the mixture without the use of large amounts of reagents.

3.7.5.2 Slag Processes

One of the earliest uses of slag for waste treatment was in the Calcilox® Process (Elnagger et al. 1977; Conner 1990). Calcilox® is fundamentally a settling or compaction process that relies on the addition of a proprietary ingredient to slowly form a cementitious mass under the overlying water layer. In normal operation, this process requires the water layer for proper compaction and reaction, and, in this respect, is much different from other stabilization/solidification processes. In some ways it is more related to conventional waste treatment than to stabilization/solidification work. It has been marketed, however, as a stabilization/solidification process and

5. Reprinted by permission of Chapman & Hall from "Chemical Fixation and Solidification of Hazardous Wastes" by Jesse R. Conner. Copyright 1990 by Chapman & Hall.

competes in certain areas with other stabilization/solidification processes. The proprietary ingredient, Calcilox®, is actually finely ground blast-furnace slag obtained as a waste product from basic steel producing plants. The process developer, Dravo Lime Co., produces Calcilox® in quantity at a plant in the Pittsburgh, Pennsylvania area. Calcilox® is typically added in the amount of about 5 to 10% by weight to the waste (Pojasek 1980). The composition of the sludge being treated is quite critical, both in solids content and as to other chemical factors. The process may use other ingredients, such as, fly ash, kiln dust, lime or portland cement, as well as the ground slag.[6]

This process has been used to stabilize flue-gas desulfurization sludges and coal waste fines in very large projects. The ground slag is mixed with the dilute waste slurry, which is then allowed to settle and compact in an impoundment where the supernatant water is drawn off and recycled or discharged. One such project has been operated at the Bruce Mansfield power station near Pittsburgh, Penn., since 1975. The capital cost for this system was more than $70,000,000. The planned capacity was 16,300 tonne (18,000 ton) of sludge per day, at a reported price of $2.75/tonne ($2.50/ton), or a daily cost of approximately $45,000; the project is scheduled to run for 30 years before the disposal site is filled. The process has applications where very large volumes of dilute waste are produced and where large areas for settling and compaction ponds can be constructed and operated over long periods.

Slag has probably been incorporated into a number of stabilization processes, along with other reagents, especially at or near the slag producers, such as steel mills. As with other waste product reagents (fly ash and kiln dusts), slag usage is often not documented in the literature or promoted specifically as a commercial stabilization/solidification process. It is used in a proprietary way at waste generators and industries. Because it is a waste product, and its composition varies considerably from generator to generator, its utility will also vary. Some of the earliest literature on the use of slag for hazardous waste treatment goes back to 1956 (Wunderly 1956), where its use for treatment of waste pickle liquor was described. Quienot (1978) described a process for using slag in the presence of an alkaline material and sulfate for solidification. Alkaline slags are known to be reac-

6. Reprinted by permission of Chapman & Hall from "Chemical Fixation and Solidification of Hazardous Wastes" by Jesse R. Conner. Copyright 1990 by Chapman & Hall.

tive in the presence of water, or can be made reactive by the addition of alkalies. This reaction probably involves the formation of soluble silica species as intermediates, followed by reprecipitation of metal silicates or calcium and/or aluminum silicates. Alternatively, slag may be made strongly acidic to accomplish the same purpose, although the reactions may be different (Rysman de Lockerente 1976). Slag is often mixed with cement, fly ash, or kiln dusts to further enhance the properties of the product (Matsushita 1978). It has reportedly been used commercially in Canada at Atlas Steel Company (OWMC 1986) for treatment of waste acid. Slag was also found to be effective in immobilizing lead (US EPA 1990a).

Blast-furnace slag is produced when the molten slag from an iron-producing blast furnace is cooled quickly to minimize crystallization. It is a blend of amorphous silicates and aluminosilicates of calcium and other bases. Because of the presence of ferrous iron and reduced sulfur compounds, it may act as a reducing agent for metal species, such as chromium, that are less mobile in the reduced valence state. It is also reactive with alkalies and acids, reportedly producing soluble silica species (Ezell and Suppa 1989).

Oak Ridge National Laboratory (ORNL) Process. The ORNL tested various mixtures of slags in combination with portland cement and fly ash for the stabilization of radioactive wastes containing technetium and nitrates (Gilliam, Dole, and McDaniel 1986). Technetium (^{99}Tc) is more mobile in the higher valence state, Tc^{+7}, than in the +4 oxidation state. Therefore, reduction of Tc^{+7} to Tc^{+4} is desirable in a stabilization process. This can be accomplished with a variety of common reducing agents such as $FeSO_4$ or Na_2S, but it was believed that blast-furnace slag, because of the presence of ferrous iron and sulfur, might accomplish the same purpose at lower cost. The purpose of this project was to test that hypothesis. The results are given in Table 5.19 (on page 5.28).

SoliRoc™. The SoliRoc™ system is really not a solidification process, but rather a complete hazardous waste treatment facility and system. While the SoliRoc™ system uses solidification as the final step, the solidification part of the process is performed using portland cement. Therefore, in this respect, it can be viewed as a cement-based process. The process consists of a fairly complex pretreatment process: generation of soluble silica species in situ at low pH, reaction of the soluble silica with metal ions, increasing pH to precipitate the system as a sludge, dewatering the sludge, and, finally, solidification with portland cement. Because of the overall approach, the process can be viewed as innovative, especially from the U.S. perspective (it has been available for some time and used commercially in Europe). Other names associated with the SoliRoc™ process are Cemstobel, S.A., and Mechim, S.A., of Belgium.

Figure 3.9 shows the process flow for the SoliRoc™ system. The pre-treatment (Stages 1 and 2) steps are:

- Formation of monosilicic acid by mixing blast-furnace slag with acidic waste (or acid) at a pH of about 1.5, wherein the silicic acid formed reacts with polyvalent metal cations; and
- "Polymerization" of the silicates to form a gel at a pH of about 11 to 12.

The resulting metal sludge can then be dewatered and disposed of in the land, or solidified with cement, lime/slag, or another generic solidification process. The main claim for the process is that metal silicates so formed are more insoluble than metal hydroxides, especially under acid leaching conditions. It is also claimed that the process is effective on certain chelated metals.

The first plant using this process was started in France in 1976 by a company called GEREP, in north Paris. It has a reported annual capacity of 10,000 tonne (11,000 ton) of waste (MECHIM 1975). As of 1984, three more plants were in operation in Belgium, Italy, and Norway.

Figure 3.9
Flow Diagram of SoliRoc™ Process

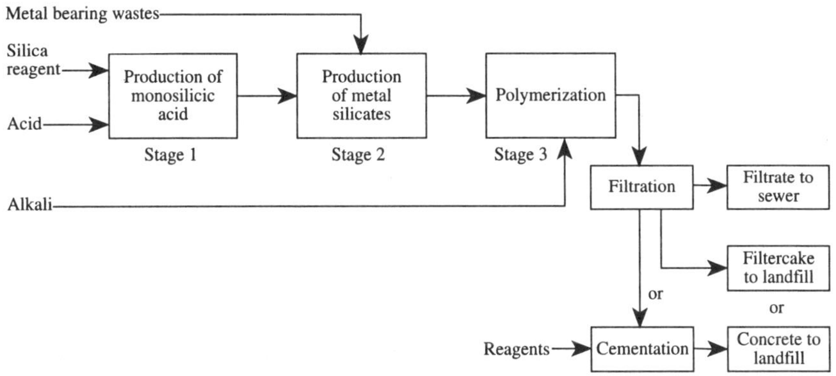

Cement-Slag Process. The combination of portland cement and slag for ordinary stabilization is not considered innovative, but its use for stabilization of hexavalent chromium is unusual and deserving of discussion here. The treatment of Cr^{+6} with slag and cement was reported by Rysman de Lockerente et al. (1976), but no leachability data were given. A laboratory treatability study was conducted on a sodium dichromate-contaminated soil, using three reducing agents to reduce the Cr^{+6} to levels that could be stabilized to within US EPA Toxicity Characteristic (TC) limits (US EPA 1990d). The agents were sodium metabisulfite and ferrous sulfate, both standard chromium reducing agents, and blast-furnace slag. When the reducing agents were used alone, neither sodium metabisulfite nor slag met the requirement, but ferrous sulfate did. When the reducing agents were combined with cement in a complete stabilization/solidification process, however, both slag and ferrous sulfate gave nearly the same acceptable results.

3.7.5.3 Lime

Processes that use lime, in general, would not be considered innovative, but one such process, DCR, has been pretreated with additives that cause it to become hydrophobic. This approach allows the improved treatment of oily and other high organic content wastes, according to the developers. The commercial offering of DCR/Boelsing/Sound Environmental Services, Inc. uses this same basic approach and the Separation and Recovery Systems (SRS)/EIF approach may be similar. The latter vendor does not disclose the exact nature of the lime that "is specially prepared and contains proprietary chemicals," but the SRS literature refers to some of the same remedial projects in Europe as does the DCR literature.

The DCR process is based on a patent by Boelsing (1977) that describes the chemistry of the method. Quicklime (CaO) is treated with a surfactant, 0.001 to 10% of its weight, that delays the reaction between the CaO and water until the CaO has first interacted with the organic matter in the waste. Subsequently, the CaO-organic mixture is converted into dispersed $Ca(OH)_2$ solid that gradually converts to limestone ($CaCO_3$) by reaction with carbon dioxide from the air. The surfactants employed and disclosed in the patent include fatty acids, paraffin oil, and aliphatic alcohols. The theory is that the high specific surface area and internal cavities of the lime will sorb large amounts of organics when the surfaces have been rendered hydrophobic. Eventually, the water penetrates the hydrophobic surface of the CaO, reacts with it in an exothermic reaction, and:

> "fractures" into powder-like material with particles in the sub-micron range. The oil which was previously adsorbed on the calcium oxide particles is now dispersed and irreversibly bound within the newly formed and highly adsorptive cavities of the hydrophobized $Ca(OH)_2$ crystals. The finely dispersed $Ca(OH)_2$ will then slowly react with natural CO_2 to generate $CaCO_3$, while still bound in the original soil matrix....The formation of insoluble calcium carbonate occurs on all exposed calcium hydroxide surfaces which are covered by a coherent carbonate crust....The specific surface area of the $CaCO_3$ decreases to less than 0.5 m^2/g as the porous surfaces are sealed by the insoluble glass-like carbonate. (Payne, McManus, and Boelsing 1991.)

This theory is supported by scanning electron microscope studies reported in the source.

The SRS process is described somewhat differently (SRS 1988a). It uses a two-step approach. First, a modified lime preparation is mixed with the waste. About 15 minutes later, a second lime preparation is added to complete the process. Thus, it appears that there is some difference between the processes, at least in the way that they are applied.

3.7.5.4 Inorganic Polymers

In soluble silicate processes, the silicates generally contain a wide range of molecular sizes from monomers to colloidal size. Depending on the actual reactions during stabilization, some polymerization may occur, but much of the gelling process is simply precipitation of the silicate species with metal ions. In at least one process, however, "Geopolymer," the primary reaction mechanism does appear to be the formation of literal inorganic polymers, based on silicon and aluminum, through chemical reaction. These materials, also called polysialates, were described by Davidovits (1982) as mineral polymers having "the empirical formula:

$$M_n\left[-(Si-O_{2-})_z-Al-O_2-\right]_n, wH_2O \qquad [3]$$

where z is 1, 2 or 3, M is sodium, or sodium plus potassium, n is the degree of polycondensation, and w has a value up to about 7. The method for making these polymers includes heating an aqueous alkali silico-aluminate mixture having an oxide-mole ratio within certain specific ranges for a time sufficient to form the polymer." In a later patent, Davidovits and Sawyer (1985) describe "an early high-strength mineral polymer composition ...formed of a polysialatesiloxo material obtained by adding a reactant mixture consisting of alumino-silicate oxide (Si_2O_5,Al_2O_3) with the aluminum cation in a four-fold coordination, strong alkalies such as sodium hydroxide and/or potassium hydroxide, water and a sodium/potassium polysilicate solution; and from 15 to 26 parts, by weight, based on the reactive mixture of the polysialatesiloxo polymer of ground blast furnace slag."

These compositions were designed to be used as high-strength cements or molded products for various commercial uses and for construction purposes, especially barrier walls, using grouts composed of geopolymer. The geopolymer binder was further described as the "result of polycondensation of a still hypothetical monomer, the orthosialate ion, producing poly(sialate) (-Si-)-Al-), poly(sialate-siloxo) (-Si--Al-O-Si-O) and poly(sialate-disiloxo) polysialates..... Geopolymeric polycondensation has also been discussed as alkali-activation of silico-aluminates by the U.S. Army Corps of Engineers." (Comrie, Runte, and Davidovits 1988).

The geopolymer system consists of two components: a very fine and dry powder and a syrupy, highly alkaline liquid (Anonymous 1989). The two are combined to produce a high viscosity mixture which is then reacted with the waste. The resulting mixture can be molded. The cured product is characterized by high strength and hardness and ability to resist chemical attack, especially by acids. The latter property results because, unlike portland cement, lime does not play a part in the lattice structure of the solid. Hazardous metals are "locked" into the three-dimensional framework of the geopolymeric matrix (Comrie, Runte, and Davidovits 1988). Therefore, the ability of a geopolymer to immobilize metals does not rely on the alkalinity of the system to maintain resistance to leaching, and, thus, the geopolymer should exhibit exceptional long-term durability even under acidic environmental conditions. This property should answer many of the commonly expressed environmental concerns about conventionally stabilized wastes.

3.8 Soluble Phosphates

3.8.1 Description

Soluble phosphates and lime have been used commercially to stabilize fly ash and mixtures containing fly ash resulting from the combustion of municipal solid waste. It has been postulated that this process may also be of use in the stabilization of other wastes heavily laden with metals, such as, medical waste ash, insulation wastes, metals smelting dusts, contaminated soils and metal contaminated sludges. It is primarily effective against lead and cadmium, but may be of benefit also in controlling other toxic metals.

The process involves the addition of various forms of phosphate and alkali for control of pH as well as for formation of complex metal molecules of low solubility. The intent is to immobilize or insolubilize the metals in the solid waste over a wide pH range. Unlike most other stabilization/solidification processes, soluble phosphate processes do not convert the waste into a solid, hardened, monolithic mass. Instead, the treated waste retains its particulate nature. It remains free-flowing, and increases little in volume.

3.8.2 Scientific Basis

The process is based on the conversion of lead or cadmium to metal phosphates of very low solubility. Table 3.5 (on page 3.46 (Conner 1990)) indicates the solubility of various phosphate species and shows that most metal phosphates are highly insoluble in water.

Many metals can be precipitated as silicates, sulfides, hydroxides, carbonates, etc. The solubilities of these species will depend, to some extent, on the complexing ion and on pH. Phosphates typically have the ability to bind with lead and other toxic metals in insoluble complexes over a relatively wide pH range, although they may resolubilize under acidic conditions.

The effect of phosphate addition on precipitation of toxic metals can be illustrated by examining its influence on speciation of lead. The dissolution or precipitation of a solid phase can be described by the solubility product,

Table 3.5
Solubility of Various Metal Phosphate Species[(a)]

Metal	Solubility
Ag	A
Al	A/L
Ba	A
Be	W
Cd	A
Ca	A
	300
Cr^{3+}	W
Co	A/L (NH_4OH)
Cu^{2+}	A
Fe^{2+}	A
Fe^{3+}	A(HCl)
Pb	A/L
	0.08
Mg	A
	200
Mn	w
Hg^{1+}	A
Hg^{2+}	A
Ni	A
Sr	A
Tl	5000
Sn^{2+}	A
Zn	A/L (NH_4OH)
	26,000

Note: W = water soluble;
w = slightly water soluble;
A = insoluble in water, soluble in acids;
L = insoluble in water, soluble in alkalies;
numbers are mg/L in water at pH 7.0
(a) Adapted from Conner 1990, 64-67.

K_{sp}. The following equation depicts a stoichiometric reaction between a metal cation, M^{m+}, and a base salt, A^{a-} (Eighmy et al. 1991).

$$M_aA_m \Leftrightarrow aM^{m^+} + mA^{a^-} \qquad [4]$$

At equilibrium, the reaction is described by the solubility product for the solid phase, M_aA_m:

$$K_{sp} = \frac{\left[M^{m^+}\right]^a \left[A^{a^-}\right]^m}{\left[M_aA_m\right]} \qquad [5]$$

Because the activity of the solid phase is considered to be unity, equation [5] can be simplified to:

$$K_{sp} = \left[M^{m^+}\right]^a \left[A^{a^-}\right]^m \qquad [6]$$

Many precipitation/dissolution reactions are influenced by participation of Lewis acid-type salts so that equation [4] becomes:

$$M_aA_m + (m \bullet a)H^+ \Leftrightarrow M^{m^+} + mH_aA \qquad [7]$$

which, at equilibrium, is described by the thermodynamic equilibrium constant, K°, so that:

$$K^o = \frac{\left[M^{m^+}\right]^a \left[H_aA\right]^m}{\left[H^+\right]^{m \bullet a}} \qquad [8]$$

Table 3.6 (on page 3.48) presents thermodynamic equilibrium constants for many lead solid phases (Lindsay 1979). These are useful in depicting the pH-dependent relative solubilities of these minerals. The more negative the log K° value, the more insoluble is the mineral. As can be seen, most lead phosphates have very low solubilities. These equations can be used to prepare solubility diagrams for the various lead solid phases, based on solution pH (see figure 3.10 (on page 3.49)). Lead concentrations below the indicated solubility isotherm for a given species reflect undersaturation and potential dissolution of a solid phase in contact with the solution. Concentrations above the isotherm reflect supersaturation and precipitation of the indicated solid phase.

As shown in figure 3.10 (on page 3.49), $Pb_5(PO_4)_3Cl$ is the most insoluble lead mineral that might form. In phosphate stabilization systems, in which high pH values result from the addition of alkali along with the phosphate, PbO and $PbSiO_3$ can also be dominant solid phases that are effective in controlling aqueous lead concentrations, while $PbCO_3$, $Pb(OH)_2$ and $PbSO_4$ tend towards greater solubility. This suggests that use of phosphate or silicate would be preferable as chemical treatments that could sequester and precipitate leachable lead in alkaline wastes.

Santillan-Medrano and Jurinak (1975) studied the solid phases controlling soil pore water aqueous lead concentrations. They found that in noncalcareous soils, the solubility of lead was controlled by $Pb(OH)_2$, $Pb_3(PO_4)_2$, $Pb_4O(PO_4)_2$ and $Pb_5(PO_4)_3OH$. In calcareous soils, $PbCO_3$ and $Pb_5(PO_4)_3OH$ were the principal controlling solid phases. The latter would

Table 3.6
Lead Solubility Equilibria[a]

Equilibrium Reaction	log K°
General Lead Solid Phase	
(1) $PbO+2H^+ \Leftrightarrow Pb^{2+}+H_2O$	12.89
(2) $Pb(OH)_2+2H^+ \Leftrightarrow Pb^{2+}2H_2O$	8.16
(3) $PbCO_3+2H^+ \Leftrightarrow Pb^{2+}+CO_2(g)+H_2O$	4.65
(4) $PbCO_3Cl_2+2H^+ \Leftrightarrow 2Pb^{2+}+CO_2(g)+H_2O+ 2Cl^-$	-1.80
(5) $Pb(CO_3)_2(OH)_2+6H^+ \Leftrightarrow 3Pb^{2+}+2CO_2(g)+ 4H_2O$	17.51
(6) $PbSO_4 \Leftrightarrow Pb^{2+}+SO_4^{2-}$	-7.79
(7) $PbSiO_3+2H^++H_2O \Leftrightarrow Pb^{2+}+H_4SiO_4$	5.94
(8) $PbS \Leftrightarrow Pb^{2+}+S^{2-}$	-27.51
Lead Phosphate Solid Phases	
(9) $Pb(H_2PO_4)_2 \Leftrightarrow Pb^{2+}+2H_2PO_4^-$	-9.85
(10) $Pb(HPO_4)+H^+ \Leftrightarrow Pb^{2+}+H_2PO_4^-$	-4.25
(11) $Pb_3(PO_4)_2+4H^+ \Leftrightarrow 3Pb^{2+}+2H_2PO_4^-$	-5.26
(12) $Pb_4O(PO_4)_2+6H \Leftrightarrow 4Pb^{2+}+2H_2PO_4^-+ H_2O$	2.24
(13) $Pb_5(PO_4)_3OH+7H^+ \Leftrightarrow 5Pb^{2+}+3H_2PO_4^-+ H_2O$	-4.14
(14) $Pb_5(PO_4)_3Br+6H^+ \Leftrightarrow 5Pb^{2+}+3H_2PO_4^-+ Br^-$	-19.49
(15) $Pb_5(PO_4)_3Cl+6H^+ \Leftrightarrow 5Pb^{2+}+3H_2PO_4^-+ Cl^-$	-25.05
(16) $Pb_5(PO_4)_3F+6H^+ \Leftrightarrow 5Pb^{2+}+3H_2PO_4^-+ F^-$	-12.98

[a] Adapted from Lindsay 1979

Figure 3.10
Solubility Diagrams for Simple Mineral Phases Containing Lead

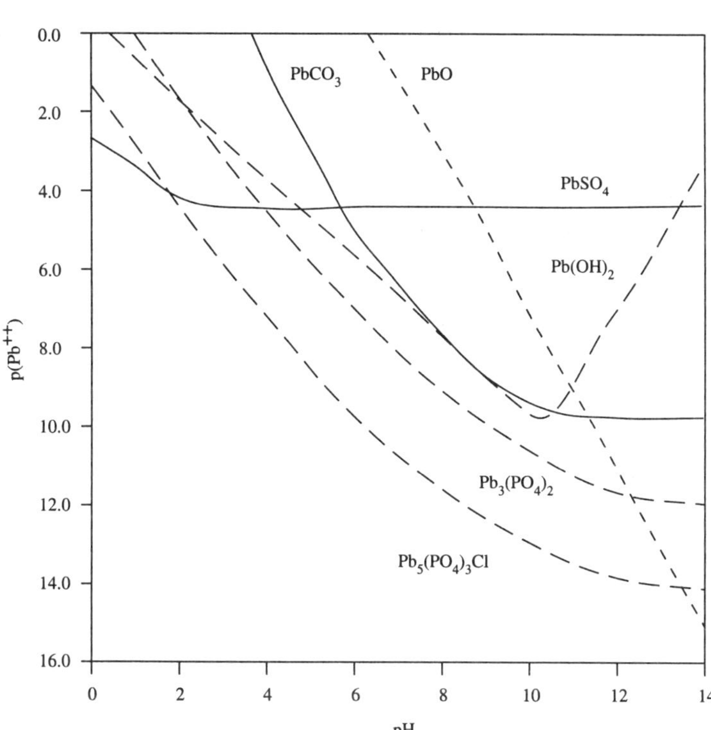

Adapted from Bobowski 1991

be analogous to a phosphate stabilization system in which lime was added along with the phosphate. These findings are in agreement with the solubility diagram in figure 3.10.

Figure 3.11 (on page 3.50), a solubility diagram for various lead phosphate minerals, (Eighmy et al. 1991), indicates that the addition of soluble phosphates to a waste containing lead will convert the lead into very insoluble species. In most cases, the solubility is below 1 x 10^{-6}M and may be as low as 1 x 10^{-14}M. Figure 3.11 shows also that the immobilization occurs

Figure 3.11
Solubility Diagrams for Lead Phosphate Minerals

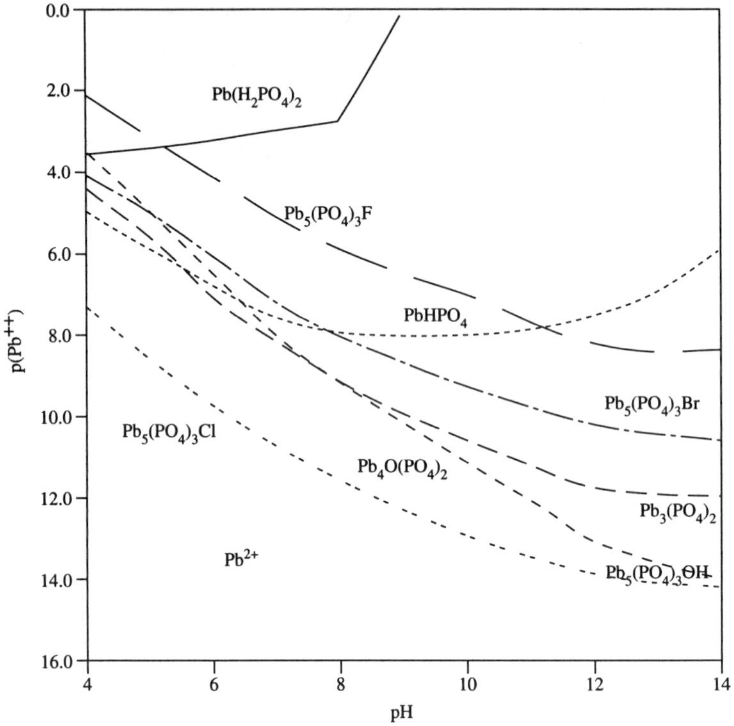

Adapted from Eighmy et al. 1991, 20

over a wide pH range. Only under highly acidic conditions will lead concentrations in leachate become important. Adding lime along with the phosphate helps assure that there is sufficient alkalinity to prevent this from occurring.

3.8.3 Status of Development

Wheelabrator Environmental Systems Inc. has a patented process (patent no. 4,737,356; April 12, 1988) called the WES-PHix process which uses soluble phosphate technology for immobilization of lead and cadmium in ash residues and other waste streams. It is a totally enclosed, in-line system which, it is claimed, reduces leaching of lead and cadmium to below TCLP limits, eliminating the need for Resource Conservation and Recovery Act (RCRA) Part B permits. The process, developed in 1985, has been permitted by several state and local regulatory agencies. The WES-PHix process technology is currently being offered by Wheelabrator to various industries internationally through site-specific license agreements. Because it is totally enclosed, potential health or safety problems for workers are minimal. The resulting product is still in particulate form, with little increase in volume, so final disposal requirements are minor.

3.8.4 Operational Considerations

Mix formulations must be determined for the particular site, although the patent provides some general guidance. The fly ash to be treated is first mixed with lime in a ratio of 0.2 parts lime to 5 parts fly ash by weight. The fly ash-lime mixture is then mixed with bottom ash, generally in the range of 5 to 20% fly ash by weight. The lime-fly ash-bottom ash mixture is then treated with water soluble phosphate to complete the immobilization. The amount of water soluble phosphate required will depend on such variables as alkalinity of the solid residue, its buffering capacity, and the amount of lead and cadmium initially present. Generally, an amount of water soluble phosphate source equivalent to between 1% and about 8% by weight of phosphoric acid, H_3PO_4, based on total solid residue, is sufficient.

Any convenient source of water soluble phosphate may be used. Such sources can be phosphoric acids, including orthophosphoric acid, hypophosphoric acid, metaphosphoric acid, and pyrophosphoric acid, as well as less acidic sources of soluble phosphate such as, monohydrogen phosphate and dihydrogen phosphate salts, and trisodium phosphate. However, calcium phosphate cannot be used, as it was found not to bind lead or cadmium.

3.9 *Comparative Costs*

All of the processes addressed in this monograph, with the exception of vitrification and organic polymer technologies, use much the same equip-

Table 3.7
Comparison of Two Stabilization/Solidification Scenarios

	Case "A"	Case "B"
Waste Type	Soil	Sludge
Contaminant/Level	Lead/400 mg/kg	Hex. Chromium/2,000 mg/kg
Waste Quantity	50,000 yd^3	1000 yd^3
Transportation	-	$40,000
Mobilization/Demobilization	$25,000	-
Treatability Study & QA/QC	$30,000	-
Excavation	$250,000	$5,000
Stabilization	$2,500,000	$150,000*
Landfill	$1,000,000	$100,000
TOTAL COST	$3,805,000	$295,000
TOTAL COST PER YD^3	$76.10	$295.00

* Includes state/local disposal tax

ment and processing techniques. Therefore, the differences in the cost of stabilization/solidification applied through these technologies will depend principally upon reagent costs and can be projected on this basis. Since stabilization/solidification embraces proven technologies, its overall costs are well established. For example, Conner (1992) gives a cost comparison of two stabilization/solidification scenarios, shown in table 3.7. Case "A" shows estimated costs for the large-scale remediation through stabilization/solidification of a low-hazard, lead-contaminated soil at a "dry" industrial plant site. Case "B" shows estimated costs for the stabilization and disposal of single truckloads of sludge contaminated with high levels of hexavalent chromium at a fixed treatment, storage, and disposal facility 200 miles from the waste site. Of the total price for the remediation (about $83/tonne ($75/ton)), $55/tonne ($50/ton) is for the stabilization operation itself. Of this $55, typical reagent cost for a cement-based process would be about $22, or about 40% of the overall stabilization price. This ratio is common for conventional processes on large-scale jobs.

4

POTENTIAL APPLICATIONS

4.1 Sorption and Surfactant Processes

Wastes amenable to sorption or surfactant processes include, but are not limited to, the following:

- oily waste;
- industrial sludges contaminated with inorganics and low levels of organics;
- soil contaminated with inorganics and low levels of organics;
- acid mine leachate and tailings; and
- radioactive liquid scintillation fluids.

Many demonstrations involving sorbents have been performed under field-scale conditions. The effectiveness of sorbents for stabilizing organic waste has not been demonstrated with this data; while reduction in leachable metals was observed.

The Silicate Technology Corporation (STC) process treated wood preservation-contaminated soils containing pentachlorophenol (PCP) with a total raw concentration ranging from 2,000 mg/kg to 8,300 mg/kg (US EPA 1992a). The hydraulic conductivity of the treated waste was on the order of 10^{-7} cm/sec and the unionized compressive strength (UCS) ranged between 1.2 to 5 MPa (170 to 720 psi). Weight loss from thermal cycling and wet/dry was less than 1%. Volume increase averaged 68%. Table 4.1 (on page 4.2) presents the leaching/extraction data. Concentrations of PCP were reduced 93 to 97% using an organic extraction test, however, the Toxicity Characteristics Leaching Procedure (TCLP) data showed an increase in PCP

Table 4.1
TCLP and Organic Extraction Test Data for Silicate Technology Corporation Process for Wood Preserving Contaminated Soils

Compound	TCLP Leaching (mg/l) Raw	TCLP Leaching (mg/l) Treated	% Reduction (Corrected)	Orgaic Extraction (mg/kg) Raw	Orgaic Extraction (mg/kg) Treated	% Reduction (Corrected)
Metals:						
Arsenic	1.8	0.86	91.6	NA	NA	ND
Chromium	0.13	0.245	-232	NA	NA	ND
Copper	3.42	0.090	95.4	NA	NA	ND
Organics:						
Pentachloro-phenol	1.5	3.42	-302	2,350	85.5	93.6

NA = Not Analyzed
ND = Not Determined
Note: % reduction adjusted for binder addition
adapted from US EPA, 1992

in the treated waste. The TCLP leaching values were reduced for arsenic and copper, but increased for chromium.

The International Waste Technologies (IWT) process and the GEOCON mixing process were combined for an in situ stabilization demonstration for treating a polychlorinated biphenyl (PCB) contaminated soil (US EPA 1990b). No overall trend in PCB changes from raw to treated waste were evident, although certain data point pairs showed differences. Results from the freeze/thaw test failed an arbitrary limit of less than 5% weight loss. The UCS averaged 2 MPa (288 psi) and the hydraulic conductivity was 10^{-7} cm/sec.

The Soliditech, Inc. process treated contaminated soil, waste filter cake material, and oily sludge from an oil company site (US EPA 1990c). Table 4.2 (on page 4.3) presents the leaching data. The TCLP leaching values for lead and arsenic decreased 85% to more than 99%; however, TCLP leaching values for barium increased. Polychlorinated biphenyls were not detected in the TCLP extracts of the raw or treated waste. Less than 1% weight loss after wet/dry and freeze/thaw was noted. The average volume increase was 22%. Hydraulic conductivity was 1^{-7} cm/sec. The UCS ranged from 2.8 to 6 MPa (390 to 860 psi).

Table 4.2
TCLP Leaching Results for Soliditech Process

Compound	Filter Cake Raw	Filter Cake Treated	Cake/Sludge Raw	Cake/Sludge Treated	Off Site Area Raw	Off Site Area Treated
Organics: (ug/L)						
Acetone	250	<210	1,000	<820	200	<160
Toluene	<2.0	<2.0	55	<24	270	<110
Arochlor 1242	<0.42	<0.45	<0.43	<0.11	<0.42	<0.44
Metals: (mg/L)						
Lead	4.3	<0.20	5.4	<0.05	0.46	<0.5
Barium	1.4	1.3	2.5	5.1	1.6	2.3
Arsenic	0.5	<0.20	0.28	0.47	0.60	0.40

Adapted from US EPA 1990a

The Hazcon process was utilized to treat contaminated soil with an oil and grease content up to 25% (US EPA 1989a). Reductions in leachable lead, as measured by the TCLP were observed. The TCLP values for the raw waste ranged from 1.5 to 52.6 mg/L, compared to a range of 0.005 to 0.40 mg/L for the treated waste. No conclusions were drawn about the effectiveness of this process for stabilizing organics. The UCS values ranged from 650 to 4,580 kg/cm^2 (220 to 1,570 lb/in^2). Hydraulic conductivity was determined to be near 10^{-7} cm/sec. Less than 1% weight loss occurred during freeze/thaw or wet/dry testing.

4.2 *Emulsified Asphalt*

The process patented by Conner (1986b) was designed for use with solid waste products, such as foundry sands, which contain undesirable waste products of metal foundry processes, such as, mold binding agents and metals. It was also found to be suitable for solidifying wastes that inhibit the chemical reactions of conventional solidification processes. Addition of emulsified asphalt to conventional portland cement solidification systems was found to substantially improve the end product by making it highly hydrophobic and substantially impermeable to water. Furthermore, no supernatant liquid was produced during the solidification process.

Asphalt fixation has also been suggested for wastes produced by paint removal, metal finishing, electroplating, and allied industries (Kulkarni and Rosencrance 1983). Microencapsulation with asphalt was found to greatly reduce leaching of metals and cyanide.

Several vendors are now marketing cold asphalt processes for encapsulation of petroleum-contaminated soils resulting from spills or leaks. In essence, the cold mix asphalt process uses an emulsifier to allow water to be combined with the asphalt, resulting in a mixture that has a viscosity low enough to be mixed with aggregate (contaminated soil). This emulsion keeps the asphalt particles separated from each other by a thin film of water. When the emulsified asphalt and aggregate have been mixed and placed, pressure from compaction breaks the water film to allow asphalt particles to come into contact with each other and the aggregate (Testa and Patton 1991). The resulting material could be used for highway paving, but the products of treating petroleum-contaminated soils would probably have lower strength and durability because of the presence of fines. This material may be suitable for low-use road base, parking lots, dust abatement, bank stabilization, storage areas, etc.

Applied Environmental Recycling Systems, Inc., (AERS) Salem, Mass., is recycling petroleum-contaminated soils into cold asphalt paving. Recycling is achieved by separating the soil components, crushing large pieces to a maximum size, recombining the components in proportions suitable for paving, and mixing the material with specific emulsions and additives. Because the process is conducted at ambient temperatures, volatilization is not a concern, and no extra energy is consumed (Anonymous 1990). Applied Environmental Recycling Systems, Inc., is focusing on petroleum contaminants, but the firm also is conducting tests on soils containing hazardous wastes, such as chlorinated hydrocarbons. The firm has also evaluated the process for use with foundry slag.

American Reclamation Corporation (AmRec) has a similar emulsified asphalt process. This process has been used to convert 12,000 tonne (13,230 ton) of oil-contaminated soil into safe, asphaltic concrete paving at a site in Worcester, Mass. (Anonymous 1991; Camougis 1990).

In a demonstration project for the U.S. Naval Civil Engineering Laboratory (NCEL), scientists from Battelle Memorial Institute demonstrated that contaminated sandblasting grit could be incorporated into a cold asphalt mix that could be used as a roadbed. A section of road has been laid using this asphalt and will be monitored for long-term performance (Nehring and Brauning 1992).

4.3 **Bituminization**

The bituminization process was developed primarily for low- and medium-level radioactive waste solutions. These solutions include: evaporator concentrates, filter sludges, and ion-exchange resin slurries. No sources are available on the use of this material for purely hazardous chemical wastes. The process has been evaluated, however, for mixed radioactive and hazardous chemical wastes (Simpson, Vidal, and Morris 1988). The report postulated potential applications of sludges from metal plating operations, laboratory sink drains, metal preparation cleanup operations, decontamination processes, and mop waters. In this process, the bitumen was designed to contain contaminants such as nitrates, cyanide, nickel, and cadmium, as well as radioactive wastes.

Although not yet widely used on a variety of waste sites, significant field data has been collected in past demonstrations and tests which are summarized in this section.

4.4 **Vitrification**

Vitrification systems are generally applicable to contaminated soil, sludges, slurries with radionuclides, heavy metals, and other inorganic contaminants, such as nitrates. Vitrification systems are also used to destroy combustible and organic contaminant materials that may coexist with inorganic contamination. Vitrification processes are generally accompanied by offgas treatment trains to remove volatile metals, particulates, and organics that evolve from the process before offgas discharge. In general, vitrification processes are able to handle a wide variety of wastes. The following are examples (US EPA 1992b):

- radioactive waste and sludges;
- soils and sediments contaminated with radionuclides, heavy metals and organics;
- incinerator ashes;
- industrial wastes and sludges;
- medical wastes;
- drummed wastes;
- shipboard wastes; and
- asbestos wastes.

Each of the vitrification processes may have special advantages in processing a particular waste. The following subsections discuss potential applications that are best suited to each vitrification process.

4.4.1 Refractory-Lined Melters

Refractory-lined melters are best suited for intermediate quantities of concentrated contaminants in either solid combustible or liquid slurry form that either have been previously collected or are currently being generated or stored. One example of a refractory-lined melter application is a vertical melter for the treatment of high-level radioactive wastes. This type of melter has been designed and constructed for operation within the Defense Waste Processing Facility at DOE's Savannah River Site. The vertical melter has also been used for several pilot-scale tests with simulated high-level radioactive wastes (Perez and Nakaoka 1986). Pilot-scale tests of a horizontal melter for Resource Conservation and Recovery Act (RCRA) wastes have also been conducted (Peters 1985). Other plans involve using a vertical melter at DOE's West Valley, New York, site to process high-level radioactive wastes. Recent efforts include designing a vertical melter for municipal solid waste incinerator ash applications (Chapman 1991). The system is also under consideration for processing medical wastes. Use of an electric arc furnace has been demonstrated for industrial and municipal waste by the American Society of Mechanical Engineers — Bureau of Mines in 1992. Seven commercial-sized vitrification plants using a water-cooled wall design have been built and are operating for municipal waste combustor ash. These systems are being used with ash as well as soils contaminated with hexavalent chromium, vanadium pentoxide, and sludge and industrial ash.

4.4.2 In Situ Vitrification (ISV)

The ISV process is generally suited for treatment of large quantities of contaminated soils or sludges or for treatment of soils and sludges that have been staged or stockpiled awaiting treatment. Restaging of contaminated soils becomes possible when the contamination is spread over a large surface area at shallow (<1 m (3 ft)) depths. Simple excavation techniques and stockpiling in the center of the site makes for more efficient operation of the process. Since a single processing unit is capable of treating approximately 23,000 m^3/yr (30,000 yd^3/yr), relatively large quantities of contaminated soil can be processed. Prudent application of the technology, however, calls for vitrification of "hot spots" in combination with other restoration technologies, such as capping, for the less concentrated areas.

To date, the ISV process has been used on three hazardous and/or radioactive sites, including a transuranic waste site and a hazardous chemical and radioactive waste site at DOE's Hanford site near Richland, Washington. The results of these tests have been published (Buelt, Timmerman, and Westik 1989; Luey et al. 1992). The process has also been demonstrated on a pilot-scale at a fire training pit at the Arnold Air Force Base in Tennessee (Timmerman and Peterson 1990). Records of Decision (RODs), supported by treatability tests, have selected ISV as the preferred treatment method for six Superfund sites in five states. These sites vary as to contaminants, having soils contaminated with PCBs, radium, pesticides, heavy metals, and radionuclides. At the time of this writing, the ISV process is being applied at the first of these six sites, the Parson's Chemical Superfund site in Grand Ledge, Michigan.

4.4.3 Thermal Vitrification

Rotary kiln vitrification, because of its large processing rates of 90 tonne/day (100 ton/day), is best-suited to large quantities of contaminated soils that can be easily excavated. This technology is also well-suited for incinerator fly ash. Because of constraints of the rotary kiln, however, particle size and foreign substances must be carefully controlled. In addition, because of limited retention data available for the more volatile contaminants, such as lead, arsenic, and cadmium, it is better suited for organic contaminants or nonvolatile inorganic contaminants, such as chromium. In one case, this technology was used upon soils contaminated with a chemical at a dye plant in Oak Creek, Wisconsin. The soils were found to be contaminated with carcinogenic residues and lead. Materials in the amount of 45,000 tonne (50,000 ton) were processed at this site (Sulzer, Albert, and Palmer 1988). This process has been used much more extensively, however, for treatment of organic contamination (McGowan and Harmon 1989).

Even though large capacity systems with equivalent soil-processing rates of 9 tonne/hr (10 ton/hr) have been used for the combustion of coal for decades, cyclone incinerators are just emerging from development as a hazardous waste vitrification process (Batdorf, Gillins, and Anderson 1992). Pilot-scale tests have been completed on hazardous dust containing lead, cadmium, and chromium. Tests have also been performed on liquid wastes, including wastewaters and carbon tetrachloride. The technology is currently being investigated as a process option for solidifying the low-level waste fraction of slurries currently stored at DOE's Hanford, Washington, installation.

4.4.4 Plasma Vitrification

The plasma processes are best suited for small concentrated quantities of slurries at relatively low throughput rates (0.45 tonne/hr (0.5 ton/hr)). The plasma arc is also well-suited to treat solid materials such as metals, glass, rubber, plastic, and filter elements. Tests with a 600 kw plasma torch have been completed on soils spiked with 15% oil (US EPA 1992b). These tests showed air to be the most satisfactory plasma gas over that of argon and argon/oxygen mixtures. Results showed destruction and removal efficiencies of between 99.99% and 99.999%. The plasma centrifugal furnace has been demonstrated in the US EPA SITE program with 1,800 kg (4,000 lb) of waste. The graphite-electrode DC arc furnace has completed a series of tests on simulated contaminated soil and combustible and metallic waste material for the DOE's Buried Waste Integrated Demonstration (BWID) technology development program (Surma et al. 1993).

4.5 Modified Sulfur Cement Process

The modified sulfur cement process was developed primarily for predried or slightly-wetted particulate wastes, such as, incinerator ash, contaminated soils, sludges, metals, and mill tailings (Kalb, Heiser, and Colombo 1991b). Waste form development and property evaluation studies have been performed also on the incorporation of evaporator bottoms and spent ion-exchange resins generated at commercially-operated nuclear facilities (Kalb and Colombo 1985).

Other applications of modified sulfur cement include barriers, storage bins, and disposal containers.

4.6 Polyethylene Extrusion Process

The polyethylene extrusion process has potential applications in the encapsulation of primary and secondary waste streams produced by waste management and environmental restoration activities. The process can be adapted for ex situ waste treatment, and it is conceivable that it can also be applied to in situ applications, although it has not yet been field-proven.

The following waste stream applications of the technology at DOE sites and facilities have been identified:

- Rocky Flats - nitrate salt wastes, sludges, ion-exchange resins, incinerator ash;
- Idaho National Engineering Laboratory - buried wastes, including over 2,300 m^3 (3,000 yd^3) of buried nitrate salt cake, incinerator ash, ion-exchange resins;
- Hanford - underground single shell tank wastes, predominately nitrate salts, sludges and ion-exchange resins;
- Oak Ridge - nitrate salts, sludges, heavy metals; and
- Savannah River - aqueous blow-down scrubber solution.

4.7 Inorganic, Cementitious Technologies of the Siliceous Category

This is the broadest category of stabilization/solidification processes from the standpoint of existing and potential applications. For the most part, potential applications are merely extensions of the conventional, well-proven processes that have been used for more than two decades on hazardous and nonhazardous wastes. The basis for such extended applications lies in the processes' capacities for (1) lowering leachability through chemical reaction, microencapsulation, reducing permeability, and improving physical durability, and (2) improving physical properties for better landfills or beneficial reuse.

4.7.1 Soluble Silicate Processes

The innovative soluble silicate processes are intended to extend applications primarily by lowering leachability through chemical reaction, microencapsulation, or reducing permeability. They may also, as is the Conner process (Conner 1985), be directed to solidify low-solids wastes. These processes apply primarily to metal-containing wastes for which the primary concern is the toxic metals, as defined by the US EPA.

4.7.1.1 EnviroGuard/ProTek/ProFix, Houston, Texas

This process was designed to be used in two waste treatment applications:

- where physical sorption is required to take up the excess water in low-solids wastes, while still producing a hardened product by chemical reaction, a requirement under the 1985 Land Disposal Restrictions; and
- where partially soluble metal species are present in the waste, which could continue to dissolve out over time, or diffuse out from porous particles. The slow, continuous generation of soluble silicate provides a reserve capacity that can re-speciate the dissolving metal as "silicates."

Because of its high sorptive capacity, porous structure, and high surface area, the process might also be applied to immobilize organics. Some ash also has sufficient carbon content – up to 5% – to potentially act in a fashion similar to activated carbon.

4.7.1.2 Lopat, Wanamassa, New Jersey

The Lopat process has conventional applications, but may be considered innovative in applications where the liquid soluble silicate is used to directly treat the waste. This step may be followed by solidification with conventional cementitious reagents, depending on the requirements of the project. Innovative use of the Lopat process lies in applications where a pretreatment with soluble silicate solutions is necessary or is preferable. This most often occurs where the waste to be treated is not finely divided and relatively homogeneous, but, instead, is very heterogeneous in physical form and often in chemical makeup. Such wastes include auto shredder fluff, incinerator bottom ash, contaminated debris, and some soils. Another innovative use lies in applications where the treatment chemicals are not physically-mixed with the waste, but are infiltrated into it. An example is the stabilization of contaminants in soil by injection or permeation of a fluid from the surface, without substantially disturbing the soil.

4.7.1.3 Other Soluble Silicate Processes

The authors do not know of applications of the possible soluble silicate formulations described in Section 3.7.5.1, since they are not presently used and are not being tested as commercial processes. The Conner process was designed to be used with low solids wastes where processing with liquid silicates alone would be too sensitive to mixing time and waste variations.

4.7.2 Slag Processes

Slag processes are especially applicable in remediating wastes from primary metal refining when practiced at the producer's site where the slag is available at little or no cost, or even with a credit for waste disposal. The Oak Ridge National Laboratory (ORNL) process was designed for the immobilization of technetium and nitrates. As it applies to reduction of mobile, higher oxidation states of metals, the demonstrated process would be applicable to Tc^{+7} and Cr^{+6}.

The same applies to the cement-slag process, In general, slag improves certain physical properties such as porosity and tortuosity, thereby improving retention of a number of species. The SoliRoc™ process is a special case, developed especially with strongly acidic, metal-containing waste solutions in mind (MECHIM 1975). It is most economical in treating this kind of waste, because the initial step requires acidifying the waste, unless it is already acidic, to dissolve the metals. Metal finishing, metal surface treatment, and steel pickling acids are good candidates. Other metal-bearing wastes can be treated – sludges, filter cakes, etc. – but, again, the waste must be acidified before treatment, requiring an extra step and cost.

4.7.3 Lime

Lime is widely used in the "stabilization" of sewage sludge (Roediger 1987), primarily for odor control and pathogen reduction. Some specific stabilization/solidification applications of lime are listed in table 4.3.

Table 4.3
Specific Stabilization/Solidification Applications of Lime

PROCESS	WASTE	REFERENCE
Lime	Steel Pickle Liquor	Sandesara 1980
Lime	Ferric chloride etching waste	Oberkrom et al. 1985
Lime + Hydrophobizing Agent	Oily waste	Boelsing 1977
Lime + Special Reagent	Hydrocarbon waste	SRS 1988a&b
Lime	Incinerator ash	Japan Patent 1988
Lime + Gypsum	Petroleum sludge	Nippon Mining 1980
Lime	Phosphoric acid residue	Schroeder et al. 1980
Lime or Lime/Fly Ash	Various	DuPont 1986

Adapted from Conner 1990

Lime has been used rather extensively in recent years in remediation projects, especially in the stabilization/solidification of oily wastes and tars (DuPont 1986). It is in these applications that the processes described in this monograph find their primary use, at least, in the sense of innovative methods.

4.7.4 Inorganic Polymers Category

The geopolymer process should, technically, be applicable to all, or almost all, of the same kinds of wastes to which cement- or pozzolan-based stabilization/solidification methods are applicable, and in the same remediation scenarios.

4.8 Soluble Phosphates

The WES-PHix process was designed specifically for treatment of lead and cadmium in refuse-to-energy ash, but its use has now been expanded for treatment of medical wastes ash, insulation wastes, metals-smelting dusts, contaminated soils and metal-contaminated sludges. Based on the solubility data given in table 3.1 (on page 3.9), the process may also be suitable for many toxic metals in addition to lead and cadmium.

PROCESS EVALUATION

5.1 *Sorption and Surfactant Processes*

5.1.1 Process Performance and Effectiveness

Sorption and surfactant stabilization/solidification technology has evolved from bench-scale studies to full-scale implementation. This section will discuss relevant research. Results of full-scale implementation are discussed in Section 4.1.

Various binders were evaluated for their ability to sorb dichloromethane as part of a solidification process for a waste oil (Wolf 1988). Table 5.1 (on page 5.2) shows that generally the inorganic solids were weak sorbents, except ground slate (the fly ash was approximately 3% organic carbon). The organic fillers and binders exhibited very high partition coefficients, with the exception of coarse-grained coke. Various combinations of these binders and fillers resulted in the leaching of only a small amount of dichloromethane into distilled water, even though the solubility of this compound is high.

The attenuative capacity of three materials (kaolin, fly ash, and sawdust) was investigated in a column testing procedure being utilized in a solidification process (Benson 1980). Table 5.2 (on page 5.3) shows the capacities of these materials as to the various metals being solidified. The sawdust capacity was believed to have been enhanced due to organic complexation reactions.

Several studies have reported upon the use of organophilic clays to remove organic compounds from water. These clays have been evaluated for their ability to stabilize a wide variety of soils contaminated with organics.

Table 5.1
Partitioning Coefficients for Organic Fillers and Binders Evaluated for Stabilizing Dichloromethane

Material	Partitioning Coefficient
Inorganic Fillers:	
Quartz	2
Infusorial Silica (0.015 to 0.040mm)	113
(0.040 to 0.063mm)	25
(0.063 to 0.200mm)	23
Ground Slate	5,200
Marl	31
Chalk	33
Clays:	
Kaolinite	88
Illite	1,320
Illite/Smectite	990
Melionite	5
Na-bentonite	132
Ca-bentonite	68
Inorganic Binders:	
Kiln Cement	29
Cement	36
Waste Kiln Dust	75
Waste Kiln Dust/Fly Ash (50/50)	14,000
Coal Fly Ash	6,100
Organic Sorbents:	
Fine Coke (<2 mm)	87,000
(0.20 to 0.63 mm)	9,100
(0.63 to 1.0 mm)	6,000
(1.0 to 1.5 mm)	5,100
Raw Brown Coal	26,000
Dry Burning Brown Coal	37,400
Brown Coal Dust	80,000

Adapted from Wolf 1988, Table 3

An evaluation of the organophilic clays conducted with four polyaromatic compounds, each with a concentration of less than 20 mg/kg, showed significant reductions in an organic extraction test (Soundararajan, Barth, and

Table 5.2
Attenuative Capacity of Various Materials for Heavy Metals

Metal	Kaolin (mg/g)	Fly Ash (mg/g)	Sawdust (mg/g)
Chromium	0.69	0.91	1.28
Copper	0.26	0.64	0.63
Cadmium	0.05	0.22	0.11
Lead	0.28	1.60	1.42
Zinc	0.13	0.51	0.30

Adapted from Benson 1980

Gibbons 1990). Furthermore, sorbent effectiveness evaluation techniques such as fourier transom infrared spectroscopy (FTIR) and differential scanning calometry (DSC), coupled with gas chromatography/mass spectroscopy (GC/MS), suggested interactions.

In another evaluation of organophilic clays, an organic waste (petroleum, 1% by weight) was stabilized with a combination of activated carbon, organoclay, and silicate. No compounds could be detected in an organic extraction test, and FTIR data suggested interactions (Frost and Carandang 1990).

In a study involving 1,000 mg/kg each of phenol, trichlorophenol (TCP), and PCP, the relative amounts of organoclay, soil, cement, and contaminant determined the immobilization effectiveness (Sell et al. 1992). The TCP and PCP appeared to be immobilized according to the Toxicity Characteristic Leaching Procedure (TCLP) test, while the phenol was not as efficiently bound.

A decreased water extraction efficiency for organic waste up to 12% was observed when modified clays were added to cement to stabilize liquid phenol and chlorinated phenols (Montgomery et al. 1988). The stabilization increased as the degree of chlorination and hydrophobicity increased.

The use of surfactants to act as dispersing agents has been evaluated for a waste containing organic material. Table 5.3 (on page 5.4) shows a significant reduction of organics using an organic extraction test.

Table 5.3
Organic Extraction Test Data from Wastech Process

Compound	Organic Extraction (μg/kg) Raw Waste	Organic Extraction (μg/kg) Treated Waste	% Reduction
Benzene	<2.0	<1.7	ND
Toluene	28.0	<1.7	>85
Trichloroethylene	1,000.0	4.8	98.9
1,3 Dichlorobenzene	33.0	2.1	85.2
Ethyl Benzene	24.0	<1.7	83.5
M/P Xylene	12.0	<1.7	>67
1,2/1,4 Dichlorobenzene	3,800.0	68.0	95.8
O-Xylene	59.0	<1.7	93.3

ND = Not determined because of small difference.
Note: % reduction adjusted for binder dilution
Adapted from Peacocke 1991, 37-66

5.1.2 Types and Amounts of By-products

The DSC, coupled with GC/MS, suggested that the interactions between the sorbent and contaminant may have produced other compounds after sorption (Soundararajan, Barth, and Gibbons 1990). Leaching data from treated waste showed a higher concentration of organic compounds than was detected in raw waste leachate (US EPA 1990a). It is not known if this increase was caused by sorbent interaction, the higher pH due to binder cement, or the conditions of the leaching test. For these reasons, a mass balance encompassing the air release pathway, binder constituents, binder dilution, and complete leachate analysis is recommended when evaluating organic stabilization techniques.

5.1.3 Cost Per Unit Volume

Table 5.4 (on page 5.5) presents a comparison of projected cost per unit quantity by several vendors that use some type of sorbent in their binders. Comparisons should not be made among the vendors because of the differing site conditions and assumptions used in the cost projection. Reagent costs make these processes slightly more costly than conventional cement processes.

Table 5.4
Cost Per Unit of Sorbent or Surfactant Stabilization/Solidification Processes

Process Reference	Cost/Unit	Assumptions	Reference
Hazcon	\$97.75 to \$205.98/ton	300 to 2,300 lb/min batch process unit	US EPA 1989, Ch.4
IWT	\$111.50 to \$194.45/ton	38,400 tons in-situ	US EPA 1990b, Ch.4
Soliditech	\$152.00/CY	5,000 CY	US EPA 1990a, Ch.4
STC	\$190.00 to \$360.00/CY	15,000 CY	US EPA 1992, Ch.4

5.2 *Emulsified Asphalt*

The emulsified asphalt process has been shown to be effective in solidifying liquid wastes, reducing solidified waste permeability, and improving the stabilization/solidification of wastes using portland cement and soluble silicates (Conner 1986b).

Kulkarni and Rosencrance (1983) reported on the use of emulsified asphalt in treating electroplating wastes containing Cd^{2+}, Cr^{3+}, Cr^{6+}, Fe^{2+}, Fe^{3+}, Pb^{2+}, Zn^{2+}, Ni^{2+}, Hg^{2+}, and CN^{1-}. A leaching study was carried out on the immobilized waste using an acetic acid/sodium acetate leachant buffered at pH 4.5 and two extractions. Results showed that the metals and cyanide leached from the fixed samples were very low in concentration; all were less than 0.5 mg/L and several were below detection limits.

Applied Environmental Recycling Systems, Inc., evaluated the use of a cold asphalt emulsion process on petroleum-contaminated soils. Using the Extraction Procedure Toxicity Test (EP Tox), AERS found that levels of total petroleum hydrocarbons and eight metals were below detection limits, except for barium, which was present in two samples at levels of less than 2 mg/L (Anonymous 1990).

The AmRec process was evaluated using the EP Tox. No petroleum hydrocarbons were detected in any extraction liquid (Camougis 1990).

Means, et al., (1991) evaluated the recycling of spent sandblasting grit into asphalt concrete. They found the product, produced at a spent grit proportion of about 7 to 10% by weight, to be very stable and suitable for use for light-traffic roadways.

The cost of operating the cold asphalt emulsion process has been estimated to be about $88 to $110/tonne ($80 to $100/ton) of petroleum-contaminated soil; it could go up to $165/tonne ($150/ton) for material with a high metal content (Anonymous 1990).

The cost of recycling sandblasting grit into asphalt concrete has been reported to be much lower than the cost of disposal. One study found the cost of continued disposal in a hazardous waste landfill to be $1,452,000/yr versus a maximum cost of $220,000/yr for the recycling option, based on an estimated annual production of 2,000 tonne (2,200 ton) of spent grit ($110/tonne ($100/ton)) (Means et al. 1991).

5.3 Bituminization

Use of the bituminization process has been limited to the treatment of low-level radioactive wastes. Westsik (1984) presents an exhaustive experiment and study on the bitumen product in comparison with cement waste forms. He determined that the leachability of radionuclides from bitumen is very low with a fractional release rate of less than 10^{-5} fraction/day. No data are available, however, for heavy metals or organics. Volatile and semi-volatile organics would not be expected to be retained, but would be removed with the water vapor. The water vapor that issues from the extruder during processing is collected for treatment.

5.4 Vitrification

Tests, demonstrations, and applications of each of these types of vitrification processes addressed here have produced a significant amount of data on the fate of various contaminants. Subsections 5.4.1 through 5.4.4, below, present data from a mass balance perspective on the retention or destruction of contaminants when treated by the several vitrification processes.

5.4.1 Refractory-Lined Melters

The overall destruction and removal efficiency for refractory-lined melters ranges from 99.7 to 99.99999% (Pacific Northwest Laboratory 1991).

The retention of inorganic elements within the refractory-lined melter is highly dependent on the element. Table 5.5 provides some typical decontamination factors within a slurry fed vertical melter (Freeman 1988). The decontamination factor (DF) is defined as the ratio of the mass of contaminant entering the melter to the mass exiting within the offgas stream. A DF of 100 would be equivalent to a 99% retention of that element. The higher the DF, the greater the amount of material retained within the glass. The more volatile elements can be captured by the offgas system, precipitated or filtered from the scrub solution, and recycled back to the melter.

5.4.2 In Situ Vitrification (ISV)

The ISV process is capable of treating organic contaminants within contaminated soils and sludges, as well as inorganic metals and radionuclides.

Table 5.5
Contaminant Decontamination Factor For a Slurry Fed Vertical Melter

Element	Average DF Total
Aluminum	22,000
Boron	100
Cadmium	9.9
Chlorine	2.9
Cesium	14
Iron	1,800
Lanthanum	2,100
Manganese	1,800
Sodium	300
Sulfur	5.5
Strontium	1,800
Tellurium	3.0
Zirconium	22,800

The reported destruction efficiency of organics (US EPA 1992b) ranges from 90% to greater than 99.99% depending on the volatility of the contaminant and the configuration within the contaminated soil. For example, the destruction efficiency of methyl ethyl ketone, a relatively volatile compound, was greater than 99% when contained within a sealed container (Koegler et al. 1989). Although the process has not yet been adequately developed for processing sealed containers safely, preliminary tests show that gases evolving from the sealed container during processing are directed through the molten glass, allowing higher destruction efficiencies. When combined with the removal efficiency of the offgas system, the total destruction and removal efficiency of organics is greater than 99.99% and can range as high as 99.99999%.

Pacific Northwest Laboratory reports retention in glass of nonvolatile metals and radionuclides, such as, chromium, strontium, plutonium, and americium, ranging from 99.99% to 99.999%. The 0.01% to 0.001% by weight that issues from the molten soil is captured by the offgas treatment system. For semivolatile metals and radionuclides such as cadmium, lead, arsenic, tellurium, and cesium, the retention ranges from 50% to 99.9% (US EPA 1992b). When combined with the offgas treatment system, the retention can range from 99.99% to 99.99999999% (Buelt and Carter 1986).

Several studies and leachability tests have been performed on the ISV waste form. For radioactive applications, some of the most relevant studies are the leachability results from the radioactive demonstrations at Hanford's 116-B-6A site (Luey et al. 1992) and the radioactive test at Oak Ridge National Laboratory (Spalding 1992). Table 5.6 summarizes the normalized release rates for given elements and contaminants.

Table 5.6
Normalized Elemental and Contaminant Releases from Radioactive ISV Products

Element/Contaminant	Normalized release over 28 day period, g/m^2	
	Hanford 116-B-6A Crib	ORNL ISV material
Aluminum	0.035	0.03
Calcium	0.015	0.02
Potassium	-	0.04
Magnesium	0.01	0.05
Silicon	0.04	0.035
Strontium	0.13	-

Spalding (1992) also indicates that the fraction of Sr-90 extracted by acid solutions was reduced by more than two orders of magnitude after treatment from 0.8 wt % to 0.006 wt %.

When scrap metals exist, it is important to determine the relative durability of the metallic phase that forms at the bottom of the vitrified product. Although no radioactive contaminant information is available, toxicity results have been reported for heavy metals as shown in table 5.7. These results provide preliminary indications of the durability of metallic waste forms that may be counted by this process when scrap metals exist with radioactive soils.

Table 5.7
TCLP Concentrations for Metallic Phase from one Bench-Scale Test

Contaminant	Initial conc. in soil μg/g	TCLP conc., metal, mg/L	Allowable conc., mg/L
Arsenic	4400	<5	5
Barium	4400	<1	100
Cadmium	4400	<1	1
Chromium	270 to 4400	<0.2 to 2.7	5
Silver	4400	<0.1	5
Lead	50	<0.1	5
Mercury	46	<0.0001	0.2

5.4.3 Thermal Vitrification

The results of using a rotary kiln vitrifier for cleanup of contaminated soils have been reported (Sulzer, Albert, and Palmer 1988). The processing rate for this unit is 14 tonne/hr (15 ton/hr) of contaminated soil. At this process rate, the burner fuel consumption was approximately 1.85 MM Btu/hr. The operating temperature was generally 980°C (1,800°F), with the vitrified discharge material at 829°C (1,525°F). At an input concentration of 400 to 800 ppm of lead in the contaminated soil, the emission rate of lead at the stack was 0.0998 g/hr (0.00022 lb/hr). This rate accounts for a minimum removal efficiency of 99.7% from the baghouse filter, meaning that up to 99.4% of the lead was either retained in the glass or collected in the spray tower scrub solution, which results in an overall retention of 99.4%. No information is available, however, on the retention of lead within the glass product relative to that collected in the spray tower and baghouse filter. These data are important for a determination whether a meaningful fraction of lead is retained in the glass as opposed to simply being removed from the soil and collected within the offgas treatment system.

Preliminary tests with cyclone vitrification tests have shown generally less effective retention characteristics for semivolatile hazardous constituents. Even though preliminary tests show 80 to 95 wt% retention of chromium, only 10 to 35 wt% of lead and 8 to 17 wt% of cadmium were retained in the vitrified slag and were entrained with the offgas after a single pass (Batdorf, Gillins, and Anderson 1992); much greater retention values are associated with electrical and plasma vitrification systems. Despite these low retention values for semivolatile constituents, preliminary tests show that the slag passes the TCLP test for metals, and that organic constituents are generally destroyed to 99.9999 wt% (six nines) efficiency.

5.4.4 Plasma Vitrification

Until recently, no information was available on retention and destruction efficiencies of plasma systems in treating specific waste contaminants. The graphite-electrode DC arc furnace tests performed for simulated soil, combustible, and metallic waste material provides data on process performance (Surma et al. 1993; Wittle et al. 1993). Although specific efficiencies are not reported, complete pyrolysis of combustible material is indicated. In these tests, cesium retention was of primary interest, which showed a retention of between 98.1 wt% and 99.12 wt%. This level of retention compares well with other vitrification processes. Durability tests with the resultant slag showed an approximate order of magnitude reduction in leachability when compared with high-level borosilicate glasses. Leachability tests over seven days for silicon showed a release rate of 0.030 to 0.080 g/m^2 compared to 0.450 g/m^2 for typical high-level borosilicate glass. For sodium, the measured release rate ranged between 0.111 and 0.557 g/m^2, compared to 1.481 g/m^2 for high-level borosilicate glass.

5.5 *Modified Sulfur Cement Process*

5.5.1 Process Performance and Effectiveness

Major efforts in the application of modified sulfur cement have been directed towards the encapsulation of incinerator fly ash containing mixed wastes generated at the Waste Experimental Reduction Facility (WERF) at Idaho National Engineering Laboratory (INEL) (Kalb, Heiser, and Colombo 1991b).

The INEL fly ash has a total activity of about 1 to 5 becquerels/gram (40 pCi/g), consisting of mixed fission products (primarily ^{137}Cs) and activation products (primarily ^{57}Co and ^{125}Sb). Elemental analyses of the ash were performed for 12 elements. Results of these analyses, expressed as percent by weight of ash, are summarized in table 5.8. Hazardous constituents include significant concentrations of Pb and Cd. Encapsulation and disposal of this ash is further complicated by the presence of highly soluble metal chloride salts (primarily zinc chloride), which create an acidic environment in the presence of moisture (pH of the ash slurry is approximately 3.8). This condition can interfere with the solidification reaction of conventional

Table 5.8
Elemental Composition of INEL Incinerator Fly Ash

Element	Weight Percentage
Zinc	36.0
Lead	7.5
Sodium	5.5
Potassium	2.8
Calcium	0.8
Copper	0.7
Iron	0.5
Cadmium	0.2
Chromium	BDL
Barium	BDL
Silver	BDL
Nickel	BDL

BDL — Below detection limits (<0.05 wt%)

solidification materials, such as portland cement. The presence of zinc, lead, sodium compounds, and chloride compounds has been shown to impede cement hydration or weaken the ultimate mechanical properties of the waste form by causing cracking or spalling (Conner 1990).

Several mixing systems were investigated for use in encapsulating fly ash in modified sulfur cement, including high shear stirrers, emulsifiers, blenders, kneaders, and double planetary orbital mixers. Based on the processing requirements for this mixture, a double planetary orbital mixer equipped with a heat-jacketed container and vacuum capability was chosen as the most appropriate means of mixing. A laboratory-scale processing system is shown in figure 5.1.

Formulation and process development work was conducted to determine the limits and ease of processibility in producing waste forms that conform to regulatory criteria. Maximum waste loadings were determined by first processing at waste loadings above the limits of workability (i.e., extremely

Figure 5.1
Bench-Scale Double Planetary Mixer for Processing Modified Sulfur Cement Waste Forms

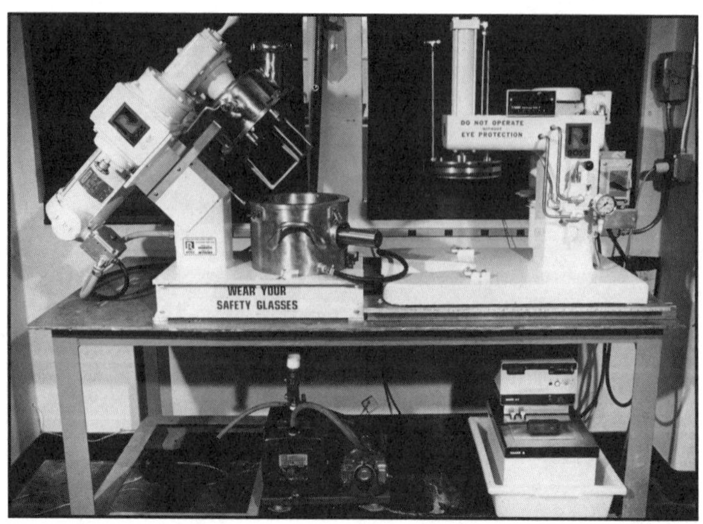

dry mixtures that yielded friable products with little structural integrity) and then adding additional increments of modified sulfur cement until acceptable workability and product integrity were achieved. Reported waste loadings represent percent by weight of dry ash, after all residual moisture has been removed. Using this procedure, a maximum waste loading of 55% by weight INEL incinerator fly ash was determined. Because of its low pH and high chlorides content, the maximum waste loading using portland cement achieved at INEL was 16% by weight.

Testing of waste form properties, including unconfined compressive strength, water immersion, freeze-thaw resistance, and leachability, was conducted on laboratory-scale specimens to provide information on structural integrity and potential waste form behavior in a disposal environment. Compressive strength data for modified sulfur cement waste forms in the range of 27.6 MPa (4,000 psi) have been reported (Kalb, Heiser, and Colombo 1991b). Additional testing to demonstrate compliance with U.S. Nuclear Regulatory Commission (NRC) criteria for low-level waste (water immersion testing, thermal cycling, and radionuclide leachability) has been reported previously for similar kinds of wastes (Kalb and Colombo 1985). The US EPA's initial test criteria for defining characteristic hazardous waste, those of the EP Tox, were recently superseded by those of the Toxicity Characteristic Leaching Procedure (TCLP). Both tests were conducted on selected formulations to assess mobility of US EPA characteristic contaminants.

Measurement of unconfined compressive strength is a general indication of a waste form's mechanical integrity and its ability to withstand loading pressures associated with overburden at a disposal site. Modified sulfur cement is a relatively brittle material and tends to fail by a shattering fracture under an axial compressive load. Thus, compressive strength testing was conducted in accordance with the standard method developed for hydraulic cements, ASTM C-39, "Compressive Strength of Cylindrical Concrete Specimens."

Results from compressive strength testing of waste form specimens containing 40 and 55% by weight INEL fly ash encapsulated in modified sulfur cement are presented graphically in figure 5.2 (on page 5.14) and are compared with compressive strength data for modified sulfur cement specimens containing no waste. Mean values for compressive strength were not highly dependent on waste loading (27.9 MPa (4,053 psi) for 40% by weight ash;

Figure 5.2
Compressive Strength Data for INEL Incinerator Fly Ash Encapsulated in Modified Sulfur Cement

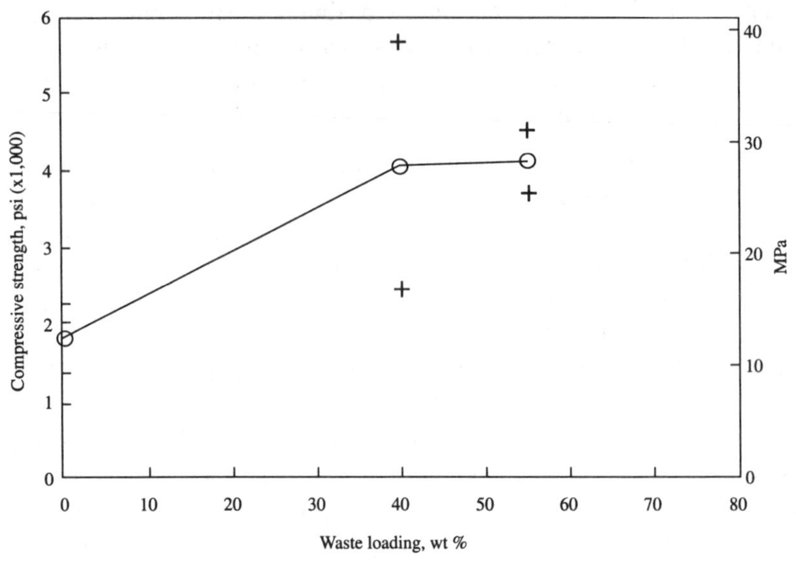

Data shown are mean values for 3 replicates (5 for 55% samples), with error bars at 95% confidence interval.

28.4 MPa (4,118 psi) for 55% by weight ash), but both waste loadings displayed strength more than twice that of the binder material alone (12.4 MPa (1,800 psi)).

Idaho National Engineering Laboratory ash and samples of encapsulated ash at several waste loadings were tested using both the EP Tox and the TCLP. Both procedures specify maximum allowable concentrations of eight metals (arsenic, barium, cadmium, chromium, lead, mercury, selenium, and silver) and a number of organic compounds. Leachate analyses were performed by atomic absorption; results are presented in table 5.9 (on page 5.15) in terms of mg/L (ppm). Although not required by US EPA, leachate concentration data are also normalized to account for the reduced mass of fly ash in encapsulated waste forms. The TCLP leachate data from

Table 5.9

Results from EPA Extraction Procedure Toxicity Test and Toxicity Characterization Leaching Procedure for INEL Ash Encapsulated in Modified Sulfur Cement

Concentrations of Criteria Metals, mg/L (ppm) [(a,b)] Sample Tested:	Cadmium	Lead
INEL Fly Ash	85.0	46.0
55 wt% Ash 45 wt % MSC[(c)] (EP Tox)	1.5 (2.7)	2.4 (4.4)
55 wt% Ash 45 wt% MSC	27.5 (50.0)	17.6 (32.0)
40 wt% Ash 60 wt% MSC	13.6 (34.0)	12.0 (30.0)
40 wt% Ash 53 wt% MSC 7 wt% Na_2S	0.1 (0.3)	1.0 (2.5)
43 wt% Ash 50 wt% MSC 7 wt% Na_2S	0.2 (0.5)	1.5 (3.5)
EPA Allowable Limit	1.0	5.0

(a) Data in parentheses represent concentrations normalized to account for reduced mass of fly ash in tested sample.
(b) Data represent TCLP results except where noted.
(c) MSC = modified sulfur cement

the INEL fly ash show that Cd (at 85 mg/L) and Pb (at 46 mg/L) are present in concentrations well above US EPA allowable limits of 1.0 mg/L and 5.0 mg/L, respectively. These elements are also evident in concentrations above allowable limits in TCLP leachates from waste encapsulated in plain modified sulfur cement, albeit at lower concentrations. (Leachate concentrations for encapsulated waste samples tested by the EP Tox method are considerably lower, demonstrating the conservative nature of the TCLP test).

Therefore, to further reduce the mobility of toxic heavy metals in the fly ash and comply with US EPA TCLP hazardous waste concentration limits, potential additives were examined, including precipitation agents, adsorption agents, and ion-exchange resins. Based on the results of scoping experiments and other considerations (e.g., ease of processing, cost, and availability) sodium sulfide was selected for use as an additive. Sodium sulfide reacts with the toxic metal salts to form metal sulfides of extremely low solubility. A ratio of sodium sulfide to fly ash of 0.175:1 was used based on the results of an experiment to determine the effectiveness of this additive on Cd mobility under US EPA TCLP leaching protocol. As seen in table 5.9 (on page 5.15), TCLP leachate concentrations for Cd and Pb (0.1 and 1.0 mg/L, respectively) from waste forms containing 40% by weight ash, 53% by weight modified sulfur cement, and 7% by weight sodium sulfide were well within allowable concentration limits. Optimization of INEL incinerator fly ash waste loading with added sodium sulfide (while maintaining additive/ash ratio constant) yielded a maximum waste loading of 43% by weight fly ash, 49.5% by weight modified sulfur cement, and 7.5% by weight sodium sulfide. As shown in figure 5.3 (on page 5.17), this represents about 2.7 times more incinerator fly ash per 55-gallon drum than is currently possible using portland cement, while still maintaining TCLP leachability below US EPA concentrations defining characteristic hazardous wastes.

5.6 *Polyethylene Extrusion Process*

5.6.1 Process Performance and Effectiveness

Polyethylene waste forms containing various forms of radioactive and hazardous wastes have been subjected to a comprehensive set of performance evaluation tests (Kalb 1993). These tests consist of those specified by the NRC for low-level radioactive waste, the US EPA for hazardous waste, the U.S. Department of Transportation (DOT) for solid oxidizers, and other testing to confirm waste-binder compatibility. Because of the lack of DOE waste form performance criteria, NRC waste form test criteria were applied. The NRC tests are stipulated in the Technical Position on Waste Form (US NRC 1991), developed in support of 10 CFR 61 (US NRC 1983). The US EPA testing for characteristic hazardous waste is defined in 40 CFR 261 (US EPA 1986). Waste form performance evaluation tests that have been conducted are listed in table 5.10 (on page 5.18), along with test

Figure 5.3

Comparison of Waste Loadings of INEL Incinerator Fly Ash in Modified Sulfur Cement and Portland Cement

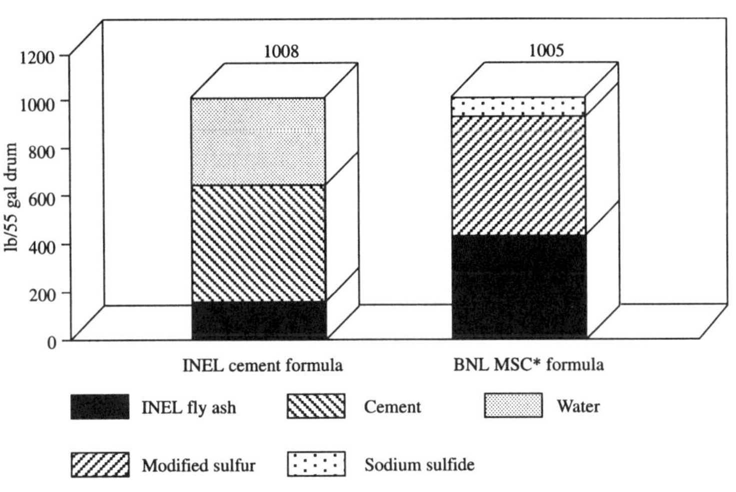

*MSC = Modified Sulfur Cement

methods and specifications where applicable. The results of these tests will be reviewed in the paragraphs immediately following in discussions concentrating on data for sodium nitrate waste encapsulated in polyethylene. Additional details on how the tests were conducted, as well as performance data for encapsulation of other waste streams in polyethylene (including sodium sulfate, boric acid, incinerator ash, and ion-exchange resins) can be found in Kalb and Colombo (1984); Franz and Colombo (1985); Franz, Heiser, and Colombo (1987); and Heiser, Franz, and Colombo (1989).

Tests were conducted using laboratory-scale waste form specimens 51 mm in diameter x 102 mm in height (2 in. x 4 in.), containing simulated sodium nitrate salt waste, or actual nitrate salt waste from the Rocky Flats Plant (RF Plant). In addition, testing was also conducted on sample cores taken from a pilot-scale (115 L (30 gal)) waste form. The pilot-scale waste form was produced during scaleup feasibility testing using a production-

Table 5.10
Waste Form Test Methods

Test	Method	Test Criteria[a]
NRC:		
Compressive strength	ASTM D-695	Compressive Strength ≥ 60 psi
90-Day Water Immersion		Compressive Strength ≥ 60 psi
Thermal Cycling	ASTM B-553	Compressive Strength ≥ 60 psi
Leachability (90 days)	ANS 16.1	Leachability Index > 6.0
Irradiation - 10^8 rad	Gamma Irradiator or Equivalent	Compressive Strength ≥ 60 psi
Biodegradation		
Fungus Attack	ASTM G-21	No observed fungal growth Compressive Strength ≥ 60 psi
Bacteria Attack	ASTM G-22	No observed bacterial growth Compressive Strength ≥ 60 psi
EPA:		
Hazardous Constituent Leachability	TCLP EP-Tox	Constituent Dependent Constituent Dependent
DOT:		
Solid Oxidizer	Test for Solid Oxidizing Substances	Comparison with reference oxidizers

(a) The minimum strength for generic waste forms specified by NRC is 60 psi. However, maximum practical compressive strengths for a given solidification agent are required. Minimum compressive strength for hydraulic cement waste forms is 500 psi.

scale 114 mm (4.5 in.) extruder to encapsulate simulated sodium nitrate salt waste in polyethylene. Results of bench- and full-scale waste form testing are given in figures 5.4 (on page 5.19), 5.5 (on page 5.20), and 5.6 (on page 5.21); and tables 5.11 (on page 5.21) and 5.12 (on page 5.22) and tables 5.13 and 5.14 (on page 5.23).

The DOT has recommended a test to quantify hazards associated with solid oxidizing materials such as nitrates. This method, "Test for Solid Oxidizing Substances," was published by the Canadian Department of Transportation of Dangerous Goods (DOT Canada 1987). It supersedes that previously recommended by DOT, "Methods for Testing for Oxidizers."

The "Test for Solid Oxidizing Substances" was conducted at Brookhaven National Laboratory (BNL) to determine effects of polyethylene encapsulation on the oxidization potential of RF Plant sodium nitrate salt waste. Results indicated that the nitrate salt solidified in polyethylene burned significantly slower (by a factor of 8 to 33 times) and less violently than any of the reference oxidizing materials (ammonium persulfate, potassium perchlorate, and potassium bromate). Compared with unsolidified sodium nitrate salt, the waste form test samples burned about 16 times slower. The solidified waste burned only about 1.3 times faster than plain polyethylene. Based on these results, sodium nitrate solidified in polyethylene is not classified as an oxidizer by the DOT, and, therefore, does not need to meet regulations for shipping oxidizers.

5.6.1.1 Economic Feasibility

This section discusses the advantages of polyethylene encapsulation of Rocky Flats nitrate salt waste from the standpoint of increased waste loading per drum, leading to a reduction in the number of processed drums required for storage, transport, and disposal. To determine net economic feasibility, savings resulting from this reduction must be balanced with overall costs, including such factors as differences in the cost of binder materials. A simple economic analysis was conducted to estimate potential

Figure 5.4

Compressive Yield Strength of Polyethylene Waste Forms Containing Sodium Nitrate Salt, Untreated and After 90 Days in Water Immersion

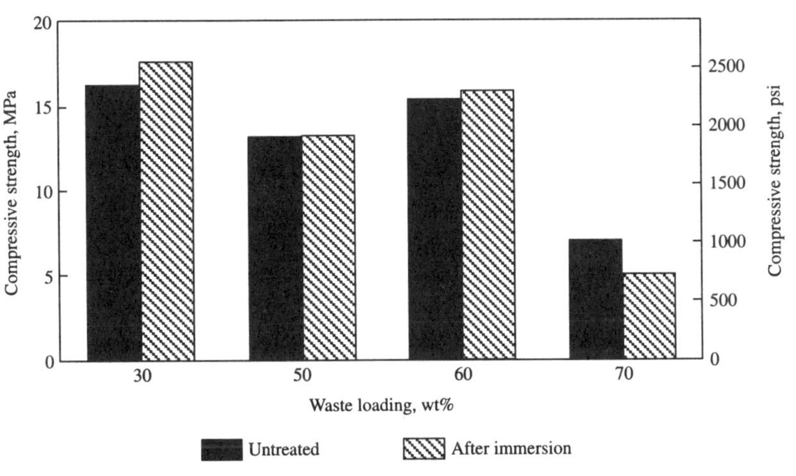

Figure 5.5
Comparison of Compressive Yield Strength vs. Waste Forms Undergoing ASTM G-21 and G-22 Biodegradation Testing and Control (Untreated) Samples

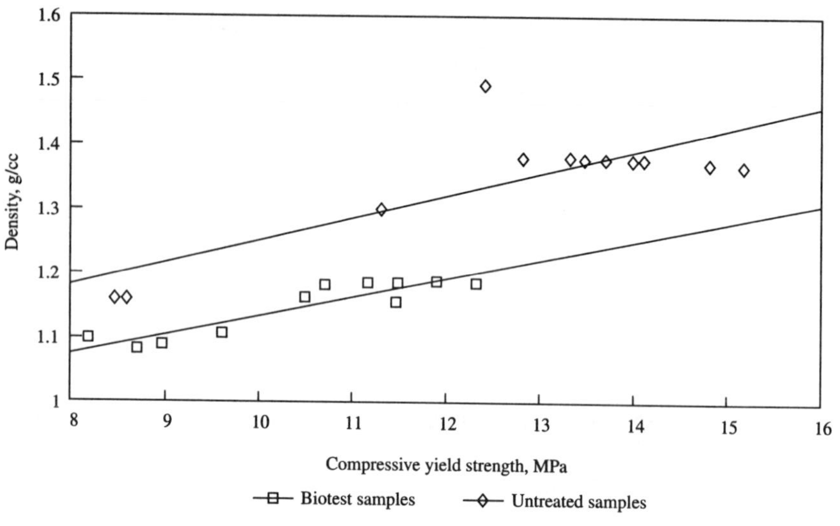

Test specimens consisted of 60 wt% sodium nitrate in polyethylene, and were cored from a pilot-scale (30 gal) waste form

cost savings from polyethylene encapsulation of nitrate salt waste at the RF Plant. A more rigorous economic analysis of polyethylene solidification of commercial reactor waste led to the conclusion that significant cost savings could be achieved using this technology (Kalb and Colombo 1985).

The analysis compares three alternatives for encapsulation of nitrate salt waste at the RF Plant:

- polyethylene;
- saltstone, (portland cement formulation using a concentrated aqueous salt solution) developed at Savannah River Plant (SRP); and
- West Valley formulation for concentrated nitrate salt encapsulation in cement.

The RF Plant has also used a portland cement-based formulation incorporating higher salt loadings (Saltcrete) developed there, but because of catastrophic failures of these waste forms in storage, this formulation was not considered (Petersen, Johnson, and Peter 1986). Maximum nitrate salt waste loadings of 13% by weight and 20% by weight in the SRP and West

Figure 5.6

Leaching Index Determined According to the ANS 16.1 Leach Test as a Function of Sodium Nitrate Waste Loading for Polyethylene Waste Forms

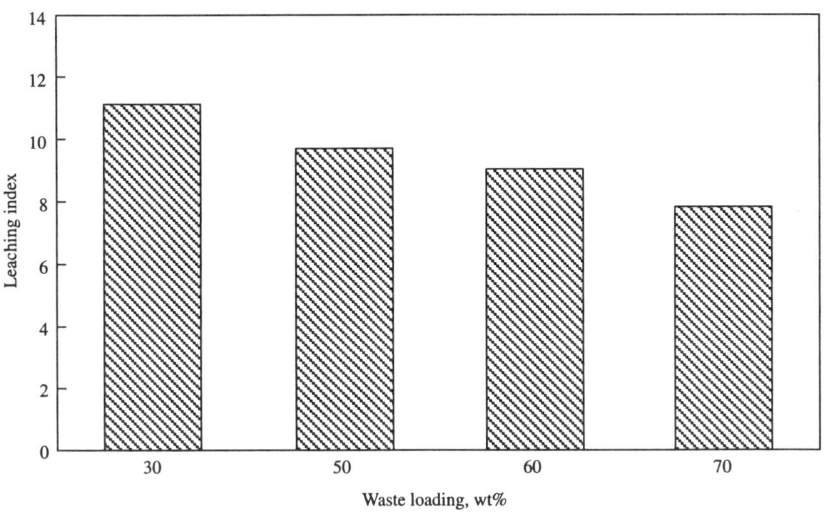

Table 5.11

ANS 16.1 Leach Test Data for Sodium Nitrate in Polyethylene Waste Forms

Weight Percent $NaNO_3$	Cumulative Fraction Leached	Leach Rate (s^{-1})	Leach Index
30	0.9	8.4×10^{-10}	11.1
50	6.3	6.0×10^{-9}	9.7
60	15.0	1.1×10^{-8}	9.0
70	73.4	1.5×10^{-7}	7.8

Table 5.12
Compressive Yield Strength of Cored Pilot-Scale Polyethylene Waste Forms Containing 60 wt% Sodium Nitrate

Test Description	Compressive Yield Strength, MPa (psi)[a,b]
Initial	14.2 ± 0.3 (2,060 ± 45)
Post Thermal Cycling	13.3 ± 0.5 (1,930 ± 70)
Post Irradiation	16.7 ± 0.7 (2,420 ± 100)
Post Biodegradation[c]	10.1 ± 1.8 (1,460 ± 255)

(a) Based on 6 replicate specimens.
(b) Error expressed as ± one standard deviation.
(c) Based on 12 replicate specimens.

Valley formulations, respectively, were taken from published literature (Wilhite 1987; McVay, Stimmel, and Marchetti 1988). Data for polyethylene waste forms were based on research and development work performed at BNL (Heiser, Franz, and Colombo 1989).

To simplify economic calculations, an annualized cost economic analysis (which considers present costs and benefits and neglects potential future variations) was performed. Some of the economic data were adapted from previous RF Plant analyses (Petersen, Johnson, and Peter 1986; Petersen, Johnson, and Swanson 1987) with allowances made for inflation. In addition, RF Plant data were based on an annual nitrate salt waste production of 800,000 kg/yr (1.76 MM lb/yr), and these calculations were updated to reflect a RF Plant production rate of 1.0 MM kg (2.2 MM lb) per year. Transportation and disposal costs are based on the assumption that waste packages will be shipped to the Nevada Test Site for disposal and on RF Plant data. Equipment costs for all systems were neglected under the assumption that these costs are roughly equivalent. Operating costs were assumed to be negligible compared to other costs and were therefore excluded. Assumptions used in the calculations are summarized in table 5.15 (on page 5.24).

Table 5.13

Self-Ignition Temperatures of Polyethylene with Nitrate Salt and Salt Waste Components [a]

Specimen	Specimen Temperature at Ignition Point, °C ± 5 °C	Air Temperature at Ignition Point, °C ± 5°C
Polyethylene	426	430
Polyethylene/50 wt% $NaNO_3$	380	381
Polyethylene/30 wt% $NaNO_3$	362	360
Polyethylene/50 wt% $NaNO_2$	360	358
Polyethylene/70 wt% $NaNO_2$	363	359
Polyethylene/50 wt% SRP Nitrate Salt	365	365

(a) Testing in accordance with ASTM D-1929, "Standard Method of Test for Ignition Property of Plastics."

Table 5.14

Results from Extraction Procedure Toxicity Test (EP Tox) and Toxicity Characterization Leaching Procedure (TCLP) for Rocky Flats Plant Nitrate Salt Encapsulated in Polyethylene[a]

	Concentrations of Criteria Metals, ppm			
Sample Tested	Chromium	Cadmium	Lead	Barium
RFP Nitrate Salt	9.0	0.4	0.5	<0.5
60 wt% RFP Salt in LDPE	3.6	0.2	0.3	<0.5
60 wt% RFP Salt in LDPE[b]	0.8	0.1	0.2	<0.5
EPA Allowable Limit	5.0	1.0	5.0	100

(a) Data for TCLP test, except where noted.
(b) Low-density polyethylene
(c) Data for EP Tox test.

Technical data used in the analysis are summarized in table 5.16 (on page 5.24), and cost breakdowns and results are included in table 5.17 (on page 5.25). Component and total costs are presented graphically in figure 5.7 (on page 5.25) and the percentages of the total cost for the BNL and SRP formulations are presented in figure 5.8 (on page 5.26). Taking into account the increased waste loading capacity and lower transport and disposal costs because of fewer and lighter packages, polyethylene represents a

Table 5.15
Assumptions Used in the Economic Analysis of Nitrate Salt Encapsulation for Rocky Flats Plant

RFP Nitrate Salt Production: kg/yr	1,000,000	
(lbs/yr)	(2,204,000)	
Materials Costs: $/kg ($/lb)		
Cement	0.22 (0.10)	
Polyethylene	0.99 (0.45)	
1986 Cost Data[(a)]		
	8.07×10^5 kg/yr (1.78×10^6 lbs/yr)	1.0×10^6 kg/yr (2.2×10^6 lbs/yr)
Labor	295,500	365,784
Repair	25,150	31,132
Miscellaneous	30,210	37,395
Waste Pre-treatment	350,860	434,311
Shipping ($/lb)	0.047	
Disposal ($/lb)	0.016	
Drums ($/drum)	28.700	
Escalation @ 5%/yr:	$1990\$=1986\$ (1+0.05)^4$	

Equipment and operating costs not included based on the assumption that:
- equipment costs for all systems are roughly equivalent
- operating costs are negligible compared to other costs

(a) Adjusted from data in reference (Petersen, Johnson, and Peter 1986; Petersen, Johnson, and Swanson 1987), to reflect increased salt production

Table 5.16
Economic Analysis for Nitrate Salt Encapsulation at Rocky Flats Plant

	Polyethylene	RFP Cement	West Valley Cement
Waste Loading (wt% dry salt)	70	13	20
Product Density, g/cm^3 (lb/ft^3)	1.67 (104)	1.70 (106)	1.65 (103)
Waste & Binder, kg/drum (lbs/drum)	329 (725)	337 (743)	328 (724)
Waste/drum, kg (lbs)	230 (508)	45 (99)	64 (141)
Binder/drum, kg (lbs)	99 (218)	292 (644)	264 (583)
Total binder, kg/yr (lbs/yr)	428,377 (944,571)	6,515,841 (14,367,429)	4,126,333 (9,098,564)
Drums/yr	4,343	22,303	15,611
Drum wt, kg/yr (lbs/yr)	49,238 (108,571)	253,144 (557,585)	176,999 (390,282)
Total Shipping wt., kg/yr (lbs/yr)	1,477,162 (3,257,143)	7,768,260 (17,129,014)	5,302,878 (11,692,846)

Table 5.17
Cost Breakdown[(a)]

	Polyethylene	RFP Cement	West Valley Cement
Labor	444,612	444,612	444,612
Repair	37,841	37,841	37,841
Miscellaneous	45,454	45,454	45,454
Portland Cement	---	1,436,743	909,856
Polyethylene	425,057	---	---
Shipping	186,077	978,560	667,998
Disposal	63,345	333,127	227,404
Drums	151,501	778,055	544,600
Waste Pretreatment	527,907	527,907	527,907
Total	1,881,795	4,582,299	3,405,673
Unit Cost, $/kg salt ($/lb salt)	1.88 (0.85)	4.58 (2.08)	3.41 (1.55)

(a) Cost data given in 1990$

Figure 5.7
Economic Analysis for Rocky Flats Plant Nitrate Salt Encapsulation

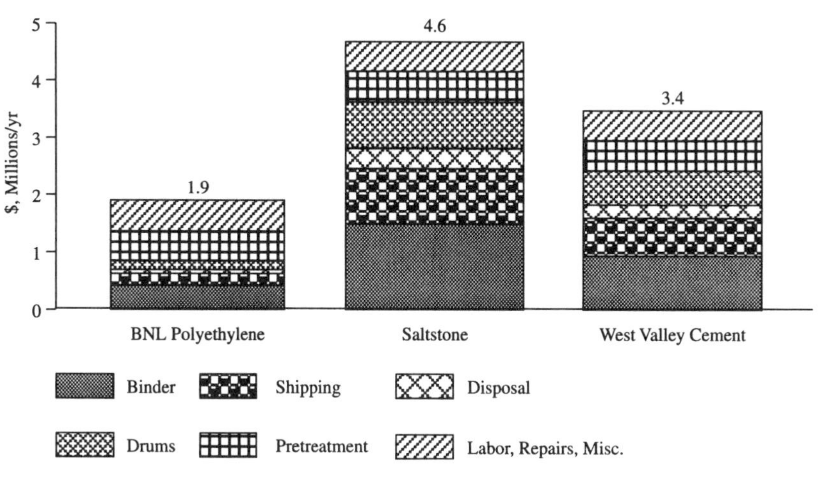

Based on RFP production of 1.0 million kg nitrate salt per year

Figure 5.8

Summary of Cost Breakdown for Economic Analysis for Nitrate Salt Waste Encapsulation at Rocky Flats Plant Using Polyethylene and Savannah River Plant Saltstone Cement Formulations

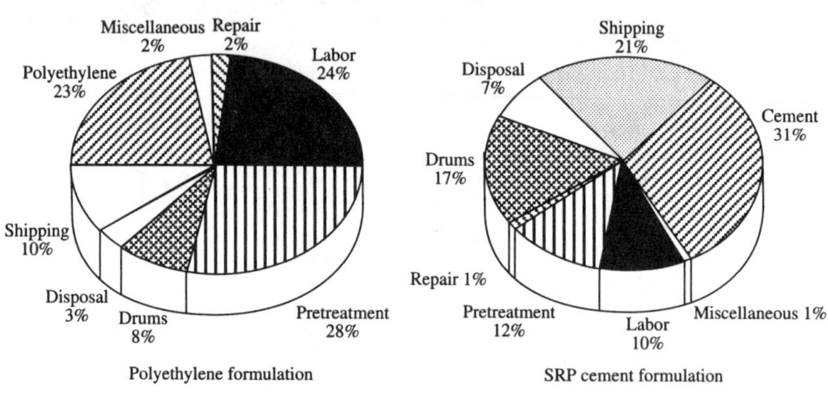

potential cost savings of about $2.7 million per year over the SRP cement formulation and $1.5 million over the West Valley cement formulation. Although these results are approximate and do not reflect rigorous economic analysis, they should provide a first order estimation of potential savings.

5.7 *Inorganic, Cementitious Technologies of the Siliceous Category*

5.7.1 Process Performance and Effectiveness

5.7.1.1 Soluble Silicate Processes

The ability of soluble silicates to reduce metal leachability by chemical means depends on their ability to react with metals to form "silicates," which are more resistant to leaching, especially to that of acidic leachants. This can occur in two ways:

- by reaction with metal ions in solution; and
- by respeciation of low-solubility metal compounds already present in the waste.

The two processes discussed below are aimed primarily at one of these two different approaches.

EnviroGuard/ProTek/ProFix, Houston, Texas. The ProFix process has been commercially applied mostly as a combined fixation agent and filter aid. The results of several uses of the stabilization system per se, however, are described by the vendor and shown in tables 5.18 and 5.19 (on page 5.28).

Data given in a patent (Conner and Reber 1992) also illustrate the hardening reactions that occur. When a mixture of water and rice hull ash was mixed with sodium hydroxide, no hardening of the paste was observed after seven days. When calcium chloride was added to the mixture, it hardened in seven days to >44,000 kg/m^2 (>4.5 ton/ft^2) bearing strength. After five months it was rock hard while the sample without the calcium chloride remained a paste. On an actual high pH calcium sludge, the addition of rice hull ash resulted in a very hard product >44,000 kg/m^2 (>4.5 ton/ft^2) bearing strength, after 12 days, although there was no measurable strength after one day. In contrast, a sample treated with sodium silicate solution demonstrated a bearing strength of 16,600 kg/m^2 (1.7 ton/ft^2) after one day, but only 17,600 kg/m^2 (1.8 ton/ft^2) after twelve days, with no additional harden-

Table 5.18
Stabilization of Organics in the Profix Process

Constituent	TCLP Leachate (mg/L)	
	Raw Sludge	Treated Sludge
Methylene chloride	20.0	<0.25
Chloroform	20.0	2.0
Trichloroethane	2.4	0.56
Toluene	2.1	0.80
Methanol	22.0	5.0
Benzene	30.0	0.76

ing thereafter. The continued hardening demonstrated by the use of rice hull ash appears to support the chemical theory of this process.

Table 5.19
Stabilization of Metals in the Profix Process

Constituent	TCLP Leachate (mg/L)	
	Raw Sludge	Treated Sludge
Lead	3.97	0.24
Chromium	7.1	0.05
Cadmium	34.8	0.07
Copper	23.5	0.43
Zinc	158.0	0.27
Nickel	32.1	0.18

Lopat, Wanamassa, New Jersey. Trezek (US EPA 1989b) described the treatment of auto shredder fluff containing about 200 mg/kg of lead with a concentrated solution of potassium silicate in water containing 1.2 to 1.9 L (0.33 to 0.5 gal) of silicate solution per ton of waste. Subsequently, the waste was treated with 10 to 12% pozzalime (lime kiln dust) and cured until dry. Leachability of lead was reduced to "acceptable" Cal WET leachable levels. The leaching levels in the untreated waste ranged from 7.8 to 600 mg/L in the Cal WET test, but <0.5 mg/L in the EP Tox. The treated waste showed leachable levels from 2.5 mg/L (untreated waste leached 7.8 mg/L) to 160 mg/L, depending on the treatment conditions. The best results reported were 6.1 mg/L in a waste that leached at the 600 mg/L level untreated. Detailed data are given by Trezek (1987) and are summarized in table 5.20 (on page 5.29) along with treatment results on other wastes. At current generic market prices, the reagent cost for this application would be only $4.40 to $5.50/tonne ($4 to $5/ton) of waste treated.

A system installed in 1987 at Hugo Neu-Proler, an auto shredder in Long Beach, California, reportedly has operated successfully (Corum 1988). The leachable Cal WET levels were acceptable to California Department of Health Services (DHS) and the facility was permitted to landfill the treated waste in a nonhazardous landfill. Other auto shredders in California have reportedly installed similar systems.

The process has also been proposed for use on lead-contaminated soils. Lead reductions of 63 to 99.97% have been reported at costs ranging from $22 to $200/tonne ($20 to $180/ton) of waste treated (US EPA 1990a). On river sediments, samples containing 320 mg/kg arsenic leached 1.0 to 1.5 mg/L as determined by the EP Tox (US EPA 1989b). A summary of other results reported by the vendor (Trezek 1987) for other wastes treated with the Lopat process are given in table 5.20 (on page 5.28). These results not only show the possible advantages of the use of soluble silicates for certain wastes, but also illustrate the care that must be exercised in their use and the variations that can occur with different additives. For example, the data on casting sand show that the use of soluble silicate alone confers no advan-

Table 5.20

Summary of Test Data on LOPAT K-20 Process

	Additive Ratio					Leachability (mg/L)				
Waste Description	K-20 gal/ton	Water gal/ton	Pz.Lime ton/ton	Cement ton/ton	Fly ash ton/ton	Metal	Untreated	Treated	Test Type	% Reduction
Incinerator ash	1.0	90.		0.2		Cd	4.3	0.4	WET	91
" "	1.0	90.		0.2		Cu	27.	38.	WET	-
" "	1.0	90.		0.2		Pb	880.	5.7	WET	99
Baghouse dust	1.0	114.	0.2			Cd	15.	<0.1	EPT	>99
" "	1.0	114.	0.2			Pb	220.	4.4	EPT	98
" "	1.0	84.				Pb	220.	<0.5	EPT	>99
Auto shredder waste	0.33	16.7		0.25	1.7	Cd	1.7	<0.2	WET	>88
" "	0.33	16.7		0.25		Pb	600.	6.1	WET	99
Contaminated soil	2.0	39.	0.2			Cr	9.3	4.4	WET	53
Battery slag	1.0	77.	0.4			Pb	740.	3.7	WET	99
Battery case residue	0.6	28.	0.34			Pb	760.	140.	WET	82
" "	0.6	28.	0.34			Pb	760.	66.	EPT	91
Casting sand	1.0	78.		0.2		Pb	160.	2.9	WET	98
" "	0.0	78.		0.2		Pb	160.	1.5	WET	99
" "	1.0	68.				Pb	160.	120	WET	25
" "	1.0	84.	0.2			Pb	160.	19.	WET	88

WET = California Waste Extraction Test
EPT = EPA Extraction Procedure Toxicity Test

tage, and at least for lead stabilization, the choice of additive (pozzalime vs. cement) makes a significant difference.

5.7.1.2 Slag Processes

Oak Ridge National Laboratory (ORNL) Process. The waste used in the ORNL test project originated in the treatment of an aqueous effluent, or "raffinate," from uranium recovery at the Portsmouth Gaseous Diffusion Plant in Portsmouth, Ohio. The baseline stabilization composition used in the test program was:

- 38.3% waste sludge;
- 11.7% water;
- 25.0% type I-II-LA portland cement; and
- 25.0% fly ash, ASTM Class F.

This yielded treated waste that leached below US EPA primary drinking water standards by the EP Tox, and resulted in Leachability Index (LI) values of about 6. For testing of the various "fixatives" – blast-furnace slag, iron filings, $FeSO_4$ and Na_2S – the filtrate was used. It had similar concentrations of technetium (Tc) and nitrate. The results are shown and compared with the treated waste without additives in table 5.21.

Table 5.21
Effect of Various Additives on Technetium Leachability

Constituent Added to Grout	Grout Composition (Weight %)					
	1	2	3	4	5	6
Raw Waste	13.9	13.9	13.9	13.9	30.0	40.0*
Water	36.1	36.1	36.1	36.1	20.0	
Cement	25.0	23.2	24.0	24.6	24.6	20.0
Fly Ash	25.0	23.2	24.0	24.6	24.6	20.0
Iron Filings		3.7				
$FeSO_4$			2.0			
Na_2S				0.9	0.9	
Slag						20.0
ANSI/ANS 16.1 Leachability Index (30 day cure)						
^{99}Tc	7.7	8.1	9.3	10.0	9.4	10.5

* Filtrate

The results clearly demonstrate the improved retention of Tc and, to a lesser extent, nitrates, by addition of granulated blast-furnace slag. Six different slag sources were tested and all gave similar results. The $FeSO_4$ and Na_2S additives gave similar results to those of the slag. Since the LI values are the negative logarithm of the effective diffusion coefficient, the slag additive improved Tc retention by three to four orders of magnitude and nitrate retention by more than one order. The Tc results are attributed largely to reduction of Tc from the +7 to the +4 valence state, accompanied by a reduction in porosity and an increase in tortuosity in the resulting matrix. The latter effect was responsible for improved nitrate retention.

SoliRoc™ Process. Data available on the SoliRoc™ process in the U.S. were most recently given by Ezell and Suppa (1989) for wastes from plants in Ohio, Michigan, and Tennessee. These are summarized in table 5.22.

The advantage of the SoliRoc™ process over that of straight lime treatment is obvious for these waste streams. Increasing the final pH resulted in little improvement in lime treatment, while the SoliRoc™ product showed little difference in leachability with final pH, except in the case of nickel.

Table 5.22
Summary of Leaching Data - SoliRoc™ Process

Formulation	Final pH	EP Toxicity Extraction Results (mg/L)			
		Cadmium	Chromium	Nickel	Lead
Ohio Waste					
Lime	9.8	6.40	0.7	21.7	0.06
Lime	12.5	5.20	0.9	8.4	0.01
SoliRoc™	9.5	0.15	0.2	1.5	0.03
SoliRoc™	11.4	0.04	0.2	1.8	0.01
Michigan Waste					
Lime	10.8	0.04	11.3	62.2	0.17
Lime	12.5	0.03	0.2	3.9	0.07
SoliRoc™	10.9	0.02	0.4	14.3	0.06
SoliRoc™	12.2	0.02	0.2	0.1	0.15
Tennessee Waste					
Lime	9.0	0.40	1.4	19.0	0.29
Lime	11.0	0.80	0.5	16.1	0.28
SoliRoc™	8.4	0.10	0.08	10.8	0.36
SoliRoc™	11.0	0.12	0.41	3.1	0.50

Cement-Slag Process. Results of the treatability study described previously are summarized in table 5.23 (US EPA 1990d).

These results clearly show that slag is most effective for chromium reduction in soils when used in combination with portland cement. In this formulation, it produced results similar to those from the standard ferrous sulfate reduction technique.

Table 5.23
Leaching of Treated, Dichromate-Contaminated Soil - Cement - Slag Process

Type of Sample	Concentration of Cr^{+6} and Total Chromium in TCLP Leachates From Treated Soil (mg/L)							
	Reducing Agent Alone				Reducing Agent + Cement (Binder/Soil Ratio = 0.2)			
	Cr^{+6}	% Reduct.	Total Cr	% Reduct.	Cr^{+6}	% Reduct.*	Total Cr	% Reduct.
Untreated Soil	38.	-	38.5	-	9.65	74.5	9.3	76
$Na_2S_2O_5$ Treated Soil	20.5	46.0	20	48.05	11.	70	9.5	75
$FeSO_4$ Treated Soil	3.65	90.35	3.3	-	1.15	97.	1.10	97
Slag Treated Soil	30.	21.05	28.5	-	2.05	94.5	1.8	95.5

*Percent leaching reduction is calculated on the basis of chromium leaching from the untreated raw soil samples.

5.7.1.3 Lime

The first large-scale commercial uses of both the SRS and DCR processes were treatment of 41,000 m^3 (54,000 yd^3) of acid tar sludge/soil at a chemical plant in Dollbergen, Germany, in 1977, and 9,000 m^3 (12,000 yd^3) of contaminated beach sand and debris from the Amoco Cadiz oil tanker spill at Brest, France, in 1978. Subsequently, these technologies were applied in projects in Japan, Canada, and elsewhere in Europe. The SRS states that 380,000 m^3 (500,000 yd^3) of waste has been treated to date with its process (SRS 1988b). Immobilization data are given for the DCR process in table 5.24.

The leachable organics have been substantially immobilized in the stabilized wastes in all cases except chlorobenzene, where the leachable amount is below detection limits even in the untreated waste. Leachable benzene in the Marathon sludge is well below the US EPA Toxicity Characteristic limit of 0.5 mg/L, despite very high leachable levels in the untreated waste.

Data provided by SRS (SRS 1988a) are of no use in evaluation, since the leachable levels for various organic constituents, even in the untreated waste, are near or below detection limits.

5.7.1.4 Inorganic Polymers

The geopolymer process has not been used commercially for the stabilization of wastes, but reportedly (Conner 1991), it has been used in Europe and Canada in fabricating construction materials. It has been tested on various wastes (Davidovits et al. 1990): mine tailings, paint sludge, arsenic-bearing wastes, and scrap yard waste. The compounds tested for environmental applications have been the sodium- and potassium-poly(sialates),

Table 5.24

Summary of Treatability Studies Using the DCR Process

Project	TPH (ppm)	TCLP Results (mg/L)				
		Benzene	Toluene	Ethyl-Benzene	Xylenes	Chloro-benzene
Marathon tank bottom sludge (dewatered)						
- before DCR treatment	10^6	368	1300	334	1830	<0.01
- after DCR treatment		<0.01	0.30	0.10	0.56	<0.01
Acid tar contaminated soils						
- before DCR treatment	$1.4x10^5$	0.21	1.10	0.38	2.4	<0.01
- after DCR treatment (8 days curing)		0.005	0.008	0.004	0.023	<0.01
Oil exploration and production sludges						
- before DCR treatment	17,700	0.526	1.69	0.334	0.479	<0.012
- after DCR treatment		<0.004	<0.004	<0.004	<0.012	<0.012
Toxicity Characteristic Limits (mg/L)	-	0.5	-	-	-	100

(sodium, potassium)-poly(sialate-siloxo), potassium-poly(sialate-siloxo) and (calcium, potassium)-poly(sialate-siloxo). Results from the treatment of various mine tailings are shown in table 5.25. The leaching test used was Ontario's Regulation 309 method, which is similar to the US EPA EP Tox. It is apparent from these data that geopolymerization is effective in reducing the mobility of the very soluble sodium and chloride ions to an unusual extent, and in other respects, in effectively stabilizing this kind of waste.

These properties take time to develop, as would be expected for a process that relies heavily on lattice substitution and/or physical microencapsulation of the hazardous constituents, i.e., on the development of a physical structure. Compressive strengths of the geopolymerized tailings were in the range of 14 to 20 MPa (2,000 to 3,000 psi) after 21 days' curing.

Similar results (Davidovits et al. 1990) are obtained in treating paint sludge waste, arsenic-bearing mine tailings, and scrap yard wastes, at geopolymer loadings of 15 to 50% by wet weight of the original waste, loadings much the same as with cement-based processes. Results in treat-

Table 5.25
Leachate Results from Geopolymerization of Mine Tailings

Waste Type	Feed (mg/L)	Regulation 309 Leachate (mg/L)		
		Untreated	Geopolymerized	
			24 Hours	28 Days
Potash waste NaCl + KCl	210,000	60,000		Na: 3,100 Cl: 7,400
Uranium waste ^{226}Ra	3,841 pCi/L	300 pCi/L		18.7 pCi/L
Metal base waste				
Iron	255,000	9,730	123	0.06
Cadmium	3.8	2.01	0.26	0.014
Cobalt	70	18.90	16.10	0.16
Chromium	756	90	45.36	<0.01
Copper	677	210	4.05	0.19
Molybdenum	2.2	0.15	0.08	0.06
Nickel	78	13.26	10.90	0.03
Lead	53	3.95	1.59	<0.02
Vanadium	119	2.4	2.20	<0.01
Zinc	1,274	802.62	484.12	3.10

ing arsenic tailings, which are thought to contain arsenic as a ferric arsenate compound, showed that the untreated waste released arsenic at the rate of about 50 mg/kg of waste, while the waste treated at the 10% geopolymer addition level released only 0.6 mg of arsenic per kg of waste in the Regulation 309 test, a reduction of eighty-fold.

5.7.2 Cost Information

With the exception of the geopolymers, all of the processes described here use reagents that are conventional and commercially-available in North America in most or many locations at competitive prices. Typical delivered prices for reagents are shown below:

REAGENT	PRICE ($/TON)
Portland cement	60 - 80
Kiln Dust	20 - 40
Blast furnace slag	10 - 20
Quicklime (untreated)	50 - 90
Rice hull ash	about $400
Sodium silicate solution	165 (FOB manufacturing plant)
Potassium silicate solution	460 (FOB manufacturing plant)
Geopolymers	880 - 1,100 (FOB manufacturing plant)

The price of hydrophobized lime in the U.S. has not been determined. Based on the ingredients, its cost should not be significantly higher than quicklime, but, as with the Lopat, geopolymer, and EnviroGuard processes, the applications are covered by patents; hence, the pricing may include appropriate markups, fees, or royalties.

5.8 *Soluble Phosphates*

The patent for the WES-PHix process presents process performance results for stabilization of several different bottom ash-fly ash-flue gas scrubber product mixtures. A number of leachants, including EP Toxicity and synthetic acid rain leachants, were used. It was found that the US EPA limits for Pb were met for untreated wastes only within the pH range 6.7 to

12.0, and, for Cd only at a pH above 7.5. In most cases, the pH of the mixtures were below these values so that the wastes would have failed the EP Tox. Treatment of these wastes with lime and either disodium hydrogen phosphate (Na_2HPO_4) or phosphoric acid greatly reduced the amount of lead and cadmium leaching and extended the range over which they passed the test to approximately pH 5.0. Table 5.26 shows the results of one such test.

Table 5.26

Reduction in Leaching Due to Phosphate Treatment[(a)]

FGSP*:Fly Ash	4:1	4:1	4:1	1:1	1:1	3:7	3:7
$\%H_3PO_4$	---	4.25	---	4.25	---	4.25	---
$\%Na_2HPO_4$			5.0		5.0	---	5.0
EP Toxicity Test							
Initial pH	12.62	12.70	12.69	6.27	12.30	5.50	11.88
Final pH	12.38	6.50	11.62	5.11	5.18	5.07	5.10
Extract, mg/L							
Lead	5.6	0.1	0.075	0.1	0.24	0.1	0.15
Cadmium	0.014	0.036	0.015	0.34	0.33	0.19	0.50

*Flue gas scrubber product.
(a)Adapted from O'Hara and Surgi(1988).

The US EPA (Kosson et al. 1991) conducted a study of leaching of a wide range of metals from ash following treatment by four processes, including soluble phosphates (table 5.27 (on page 5.37)). The Availability Leach Test results are reported here. This test is used to assess the maximum amount of specific elements or species that could be released under an assumed "worst case" environmental scenario. Two serial extractions are carried out on crushed samples. The pH of the leachant is automatically controlled to pH 7 during the first extraction and to pH 4 during the second extraction, using nitric acid. The first and second extracts are combined for analysis (van der Sloot, Piepers, and Kok 1984). This test generally extracts all species that are not tightly bound in a mineral or glassy matrix. Compared with portland cement treatment, the use of WES-PHix was found

Table 5.27

Comparison of Species Release for Availability Leaching Tests on Untreated and Treated Municipal Solid Waste Residues (mg released/kg dry ash)

	Process 1	Process 2	Process 3	Process 4	Portland cement control	Untreated residue	Total residue analysis
Bottom Ash							
Cadmium	8	10	16	17	18	28	36
Chromium	4	10	9	< 1	8	13	200
Copper	77	270	200	230	150	360	2,100
Lead	130	520	360	350	180	2,700	NA
Zinc	670	7,100	1,600	1,800	2,100	2,800	4,800
Air Pollution Control Residue							
Cadmium	220	170	220	120	150	130	200
Chromium	10	20	17	< 1	37	6	41
Copper	160	280	390	130	520	190	360
Lead	1,200	2,100	2,300	4	3,700	980	3,000
Zinc	6,700	11,000	12,000	6,700	16,000	7,700	3,000
Combined Ash							
Cadmium	32	20	20	26	27	27	36
Chromium	5	16	19	< 1	6	3	200
Copper	240	400	390	240	360	380	2,100
Lead	260	490	1,400	46	370	500	1,600
Zinc	1,700	2,500	2,000	2,200	2,300	2,900	4,800

Process 1 - Portland cement and a polymeric additive.
Process 2 - Portland cement and soluble silicates.
Process 3 - Quality controlled waste pozzolans.
Process 4 - Soluble phosphate.
Source: (Kosson et al. 1991, 127-129).

to be much more effective for Zn, Cu, Cr, and Pb control and similar for Cd control. The control of lead with soluble phosphates was found to be six times more effective than portland cement when treating bottom ash and 900 times more effective when treating air pollution control (APC) residues. Control of lead was also much better with the soluble phosphate process than with either the portland cement plus polymer or the portland cement plus soluble silicate processes.

6
LIMITATIONS

6.1 Sorption and Surfactant Processes

Reliability of the sorption and surfactant processes for organic waste cannot be assessed because field-scale monitoring of completed projects is in its infancy. Remediation personnel need to rely on predictive methodologies, such as accelerated-aging testing, to determine the long-term effectiveness of sorption processes.

6.1.1 Site Considerations

Stabilized/solidified material that is disposed of in a nonlined area has the potential to be contacted with and degraded by groundwater. The sorbed organics could become mobile if the groundwater contains organic compounds.

6.1.2 Waste Matrix and Risk Considerations

The effectiveness of all sorbents is limited to the extent that waste material is sorbed and controlled in an equilibrium condition. There is always a risk that the waste may be released, since it continues to reside within the stabilized/solidified material. The degree of risk depends on the extent leaching is controlled, which is determined by the extent the material is bound or dispersed.

6.2 Emulsified Asphalt

The emulsified asphalt process has not yet been evaluated for application to a wide range of wastes. More research is needed before the long-term durability of this material can be assessed.

Certain constituents of the waste may interfere with the production of high-quality asphalt. For example, high-organic content is detrimental to stability and sulfate may cause swelling upon contact with water.

If the waste materials are hazardous, compliance with transportation and storage regulations is required. Reuse as pavement material may be problematic.

6.3 Bituminization

The primary limitation of this process results from lack of experience and data beyond low- and medium-level waste applications. Consequently, its applicability and limitations in hazardous chemical waste areas are difficult to determine. As stated before, however, the waste solids being processed are generally in a slurry form. Solids content is usually concentrated to 50% by weight solids, the remainder is moisture. The final product is composed of approximately 40% solids and 60% bitumen.

6.4 Vitrification

Limitations of vitrification processes have been evaluated and summarized (US EPA 1992b). The following factors may limit the overall effectiveness or cost-effectiveness of vitrification:

- feed moisture content;
- feed material composition;
- feed compatibility;
- combustible material;

- potential shorting caused by metals, and
- cost of energy.

6.4.1 Feed Moisture Content

The moisture in the feed increases the amount of energy needed for vitrification. To evaporate a given mass of water generally requires about the same amount of energy as the vitrification of a given mass of solid materials at 1,200°C (2,200°F) (approximately 0.7 kwh/kg). Nevertheless, moisture limits the application of vitrification only in certain circumstances.

For example, special engineering measures may be needed to apply ISV to contaminated soil within shallow, permeable aquifers. Although ISV has been used to process sludges with moisture contents as high as 55% by weight (Buelt and Freim 1986), the potential for moisture recharge in a permeable aquifer will prevent the process from proceeding downward. Calculations show that soil permeabilities greater than 10^{-4} to 10^{-5} cm/sec will inhibit downward vitrification (Buelt et al. 1987). Where this occurs, groundwater diversion or pumping techniques would have to be implemented.

6.4.2 Feed Material Composition

One of the advantages of vitrification processes is that the feed can be augmented by glass formers, modifiers, or fluxes to achieve optimal product durability and process performance. Consequently, feed materials for refractory-lined melters, as well as water-cooled melters, thermal vitrification, and plasma systems can be augmented to eliminate any limitations associated with the feed materials' composition. For the ISV process, however, the soil must be amenable to vitrification as it exists. Most soils have the appropriate processing and product durability characteristics (Buelt et al. 1987). One of the few exceptions occurs in weathered soils that have been discovered along the eastern seaboard of the United States. The alkaline content of such soils is lower than the minimum 1.4% by weight requirement. In this case, alkaline chemicals, such as soda ash, could be premixed with the soil by deep soil mixing techniques before vitrification.

6.4.3 Feed Compatibility

Feed compatibility refers to the compatibility of the feed material as to size and nature of the material being treated. Size is generally dictated by the feed mechanisms, and for the above ground vitrification processes, the feed must be pretreated to assure that materials are of a size that can be handled. The ISV process has been shown, however, to be capable of processing large rocks and boulders, large combustible materials, such as timber, and large chunks of metal. The ISV process is not yet sufficiently developed to process buried, sealed containers, such as 55-gallon drums. These items can result in sudden release of offgases, expelling molten soil into the offgas collection hood and resulting in a loss of offgas containment.

6.4.4 Combustible Material

Combustible material generates gases of decomposition and increases the thermal heat load of the offgas treatment system. Above-ground melters are well-suited to handling combustibles because the residence time in the melters can be controlled by feed rate. The only concern is that the combustibles be compatible with the feed delivery system. Combustible feed is less frequently processed by refractory-lined melters than by the rotary kiln vitrification process, because of the combustion capacity of the rotary kiln, an extension of incineration technology. Limits on combustible feed are prescribed for ISV in order to avoid excessive heat loads in the hood and offgas system. Heat from combustion is generated above the molten glass, where pyrolyzed gases meet oxygen. Much of the heat generated is not returned to the glass and, therefore, must be removed from the offgas system. Current combustible content limits have been established at 7% by weight maximum in the soil being vitrified (Buelt 1992). The process has been successfully applied at full scale for combustible wastes, handling eighty, 3.7-m (12 ft) long, creosoted timbers in a single ISV setting. Gaseous effluents were completely contained during this operation, and the offgas system was effective in handling the increased heat load.

6.4.5 Potential Shorting Caused by Metals

When metals are introduced in a vitrification process, they can form a dense phase of molten metal at the bottom of the molten glass pool. This can cause an electrical shorting in some kinds of vitrification processes (or more rapid corrosion of refractories in others). Vertical refractory-lined melters and ISV are most suitable for handling metals. As molten metals form in a vertical melter, they can be tapped through the bottom drain into the receiving canister. The ISV technology has recently been upgraded to allow processing of wastes with high concentrations (up to 25% by weight) of metals (Buelt 1992). The electrode feeding system of ISV allows the vertical position to be controlled to avoid electrical shorting. This system has been effectively demonstrated on a large scale and is ready for field deployment. Horizontal refractory-lined melters and rotary kiln vitrification are less suitable for treating wastes containing metals because of concerns about refractory longevity.

6.4.6 Cost of Energy

Although energy costs are a significant part of the overall operational cost (i.e., up to 40% of that of ISV), the amount of energy needed to perform vitrification is generally overestimated. Energy requirements are generally 0.6 to 1.0 kwh/kg for electrical vitrification processes, such as ISV and refractory-lined melters, and up to 1.6 kwh/kg for fossil fuel fired processes, such as the rotary kiln (Buelt 1992). A study showed that less energy is required to vitrify a given volume of contaminated soil by ISV than to transport it 500 miles or more. Most vitrification processes consume between 500 and 3,500 kw. This is much less power than a large hotel would consume. Nevertheless, vitrification does require significant amounts of electrical energy. If it is not available at the site, it may have to be provided by portable generators; generator power is usually more expensive than line power.

6.5 *Modified Sulfur Cement Process*

The modified sulfur cement solidification process has been shown to be highly reliable (Kalb and Colombo 1985). Following are the primary requirements for processing:

- suitable thermal input to supply latent heat of fission and maintain a molten condition; and
- ability to thoroughly mix waste and binder under viscous conditions to form a homogenous mixture.

Generally, the waste should be predried, since sulfur is not compatible with water. Some moisture, however, can be tolerated; it will subsequently evaporate under processing temperatures. Since sulfur cement is essentially inert at temperatures used for processing, no interaction between waste and binder is anticipated.

6.6 Polyethylene Extrusion Process

The reliability of the Polyethylene Extrusion Process has been demonstrated through application to waste streams having a broad range of chemical and physical properties. The inert characteristics of polyethylene and its compatibility with many kinds of wastes over wide concentrations assures reliability of performance and predictable end products. The process, however, requires that the waste be dried, since polyethylene is not compatible with water or other aqueous solutions. Drying of wastes before their incorporation with polyethylene results in volume reduction, higher waste loadings, and a homogenous product. Confirmatory tests (Franz, Heizer, and Colombo 1987) have shown that no reactions would occur that could adversely affect either the health and safety of workers during processing operations or compromise the integrity of the end product.

6.7 Inorganic, Cementitious Technologies of the Siliceous Category

The innovative aspects of these processes are generally limited to specific applications, discussed in Subsections 6.7.1 through 6.7.4, below. The reagents used are widely available (except for rice hull ash) at competitive prices. In general, more independent evaluation is needed to assess the effectiveness and practicality of these innovations versus the conventional cementitious, siliceous technologies.

6.7.1 Soluble Silicate Processes

Soluble silicate processes are generally not effective in immobilizing organics and are not marketed for that use. There is little reason to believe that they, as a group, enter into chemical reactions with organics, as a group. As with other processes, however, there could be advantageous reactions with specific hazardous compounds. The main limitation in the use of soluble silicate processes results from their sensitivity to such operational factors as order of addition, mixing type, duration and degree, and quantity added. The latter aspect is discussed in Section 3.7.5.1.

6.7.1.1 EnviroGuard/ProTek/ProFix, Houston, Texas

The ProFix process has limited use in solidification of high-solids content wastes, such as, sludges, filter cakes, and contaminated soils, since sorbents are not required or recommended. The chief operational disadvantage lies in the cost of shipping and handling rice hull ash. This material, while inexpensive at the source, is available in only a few locations and is expensive to transport because of its low-bulk density, which also requires that it receive specialized handling at remedial sites in order to avoid dust problems and to effect proper metering into the treatment system.

6.7.1.2 Lopat, Wanamassa, New Jersey

The Lopat process, in the innovative sense, applies primarily to wastes that contain soluble toxic metal salts, but are otherwise nonhazardous. Examples include auto shredder fluff, certain contaminated soils, and various ashes. In this case, the silicate can react directly with the metal salt, forming relatively immobile metal "silicates." The process can, of course, be applied, along with cementitious agents, in conventional use treating other wastes. The innovative aspect of the process, discussed in Section 3.7.5.1, is limited to certain applications. There appears to be no data available from independent evaluations of the Lopat process. All available data came from Lopat itself, its licensees, or customers.

6.7.1.3 Other Soluble Silicate Processes

Except for the Conner process, it appears that the soluble silicate formulations described in Subsection 3.7.5.1, are not currently being used or tested as commercial processes. In any event, their use would likely be limited to specific kinds of wastes or site scenarios. The Conner process's usefulness, in the innovative sense, is limited to low-solids wastes where processing with liquid silicates alone would be too sensitive to mixing time and waste variations.

6.7.2 Slag Processes

For remediation projects, as the distance from the source to the site increases, the cost-effectiveness of slag processes decreases because of competition from locally available materials – fly ash, kiln dust, etc. The process has been applied only to metal-bearing wastes. Slag should have general applicability except where its reductive properties would be undesirable, such as in the treatment of certain arsenic-contaminated wastes. The SoliRoc™ process is highly waste specific. The economics favor highly acidic wastes where no additional cost for acid is incurred and the waste must be neutralized before solidification in any event.

6.7.3 Lime

The processes described in this monograph are designed to treat wastes containing high levels of organics. Lime processes would be considered innovative in treating metal wastes, but there appears to be no theoretical or practical advantage in such an application.

6.7.4 Inorganic Polymers Category

There are several limitations attending the geopolymer process. The most important is cost. Costs for large quantities of the reagents in Europe are in the range of $800 to $1,000/tonne ($730 to $900/ton). At present, there is no source of supply in North America. Reagent cost is about $100 to $600/tonne ($90 to $550/ton) of waste treated - typically $100 to $200/tonne ($90 to $180/ton) of waste for the examples tested. Therefore, at present, this process would be limited to waste and remediation scenarios that cannot be handled effectively by other processes (Conner 1991). Additional use may be expected if the cost of the geopolymers were reduced.

6.8 Soluble Phosphates

To date, soluble phosphates have been thoroughly evaluated only for use with municipal fly ash. Further research is needed to determine whether the process will be applicable to other wastes. Not all metal phosphates are insoluble; therefore, application is limited to wastes containing mixtures of contaminants. Organics would probably not be stabilized by this process.

7

TECHNOLOGY PROGNOSIS

7.1 Sorption and Surfactant Process Aspects Needing Further Development and Demonstration

Other natural or synthetic materials that result in strong interactions with heavy metals or organics need to be evaluated. The saturation limits of sorbents need to be evaluated for loading purposes. The degree of sorption must be assessed under realistic disposal environments, as opposed to regulatory bench-scale testing methods.

7.2 Emulsified Asphalt

The future is promising for use of the emulsified asphalt process with petroleum-contaminated soils, such as those resulting from oil spills. The product is very similar to commercially-available, low-strength asphaltic construction materials, and the cost of treatment is relatively low. Use in treating wastes containing metals or organics, such as chlorinated hydrocarbons, however, is problematic, and this application must undergo much more analysis.

7.3 Bituminization

In recent years, very little information about bituminization has been published. Consequently, the prognosis for use of this technology in treating hazardous chemical waste is quite poor. No data are available on the durability of bituminized waste forms to use in determining whether use of the product would be in compliance with US EPA restrictions on leachability, such as the Toxicity Characteristic Leaching Procedure.

7.4 Vitrification

Significant programs are continuing the development of the refractory-lined melter, melters with water-cooled walls, in situ vitrification, and plasma vitrification, for expanded applications. Several private industrial firms, government agencies, and national laboratories are developing and marketing these types of vitrification systems for the treatment of a broad variety of wastes. Several of these melters are emerging from both foreign and domestic manufacturing and municipal waste treatment industries for application to hazardous and radioactive wastes. These relatively new and emerging technologies hold a great deal of promise in the hazardous waste treatment field. Application is expected to extend beyond contaminated soil to drummed waste, tanked waste, and even reactive materials. Rotary kiln vitrification, on the other hand, is relatively mature and is generally being pursued only for contaminated soil applications.

7.5 Modified Sulfur Cement Process

The waste loading or packing efficiency of the system is dependent on the size and electrical capacity of the unit, since the material becomes more viscous with the addition of waste particulates. A large-scale demonstration is needed to determine optimum waste loadings. Further work is needed with sludges, soils, and other waste streams to better evaluate the end products.

7.6 Polyethylene Extrusion Process

Although the process has been thoroughly studied and evaluated in the laboratory, a demonstration is needed to ascertain its reliability under scale-up conditions. Planning has been initiated to demonstrate this system using production-scale equipment at Brookhaven National Laboratory.

Additional work is necessary to determine the types and capacities of dryers suitable for the removal of moisture from aqueous, wet-solid wastes before processing. Presently, spray-dryers, vacuum-dryers and thin-film evaporators are being evaluated for a wide variety of wet-solid waste streams.

Although the waste forms have demonstrated their ability to retain toxic mixed waste components, studies are needed to determine maximum concentrations of heavy metals that can be incorporated into polyethylene in compliance with regulatory requirements.

7.7 Inorganic, Cementitious Technologies of the Siliceous Category

7.7.1 Soluble Silicate Processes

7.7.1.1 EnviroGuard/ProTek/ProFix, Houston, Texas

The possible advantages of in situ generation of soluble silicate have not been fully explored. Work should concentrate on the development of long-term properties, i.e., long curing periods or long-term leaching procedures, since that is where the advantages will lie.

7.7.1.2 Lopat, Wanamassa, New Jersey

There is a need for more information on the comparative properties of different soluble silicates. There are many grades and types of alkali metal silicates whose differing alkalinities and silica contents may be important. This approach to stabilization has probably not been optimized overall.

7.7.1.3 Other Soluble Silicate Processes

There is need for further general work in this area. Various silicate setting agents and other additives may result in improved stabilization methods. This is important in view of recent, more exacting regulations and remedial project specifications.

7.7.2 Slag Processes

Broader testing of blast-furnace slag, both as a primary stabilization agent and as an additive, would be beneficial. Slag is inexpensive and available in most industrial locations. It should be compared with other additives, such as fly ash and kiln dusts, in this respect. Like these other waste materials, the properties are quite variable and slag usage requires good quality control procedures.

7.7.3 Lime

The processes described have not been directly compared with other treatment methods, or even to untreated lime, either CaO or $Ca(OH)_2$. Such a study needs to be done to establish whether the presumed additional cost of the treated lime is justified.

7.7.4 Inorganic Polymers

Relatively little test information is available on inorganic polymer processes. Especially needed are data on difficult-to-treat wastes, such as those containing significant amounts of organo-metallics, organics, and radioactive species. In remedial applications where high early strength is needed and where acid leaching conditions are expected, geopolymer use is a possibility. In these applications, the higher cost of the reagents could be justified.

7.8 *Soluble Phosphates*

Soluble phosphate treatment of fly ash from municipal solid waste incineration should continue to grow in use. Its use in treating other metals-bearing wastes, such as metals smelting dusts or contaminated soils, must be extensively tested and evaluated.

APPENDIX A

List Of References

Anonymous. 1989. Geopolymers. In *Proc. 3rd International Conf. on New Frontiers for Haz. Waste Mgt.* Pittsburgh.

Anonymous. 1990. Cold asphalt emulsion technology offered to recycle petroleum-soil. *HazTECH News* March:75.

Anonymous. 1991. Remediation saves Worcester site. *New England Waste Resources*. March:12-3.

Batdorf, J., R. Gillins, and G.L. Anderson. 1992. *Assessment of selected furnace technologies for RWMC waste*. EGG-WTD-10036. Idaho National Engineering Laboratory. Idaho Falls, Idaho.

Benson, R. 1980. Natural fixing materials for the containment of heavy metals in landfills. In *Proc. of Twelfth Mid-Atlantic Waste Conference*. Lewisburg, Penn.

Bobouski, S.F. 1991. *The effect of soluble phosphorus on the leachability of lead and other metals in residues from municipal solid waste incinerator.* MS Thesis, University of New Hampshire. Durham, N.H.

Boelsing, F. 1977. U.S. Patent 4,018,679. April 19.

Buelt, J.L. 1992. *The in situ vitrification program: focusing an innovative solution on environmental restoration needs*. PNL-SA-20853. Richland Wash.: Pacific Northwest Laboratory.

Buelt, J.L. and J.G. Carter. 1986. *In situ vitrification large-scale operational acceptance test analysis*. Richland Wash.: Pacific Northwest Laboratory.

Buelt, J.L. and S.T. Freim. 1986. *Demonstration of in situ vitrification for volume reduction of zirconia/lime sludges*. Richland Wash.: Pacific Northwest Laboratory.

Buelt, J.L., C.L. Timmerman, and J.H. Westsik. 1989. In situ vitrification: test results for a contaminated soil-melting process. PNL-SA-15767-1. Pacific Northwest Laboratory. Richland, Wash.

Buelt, J.L., C.L. Timmerman, K.H. Oma, V.F. Fitzpatrick, and J.G. Carter. 1987. *In situ vitrification on transuranic waste: an updated systems evaluation and applications assessment.* PNL-4800 Suppl. 1. Richland Wash.: Pacific Northwest Laboratory.

Bikales, N.M. 1967. Ethylene polymers. Vol. 6. *Encyclopedia of Polymer Science and Technology*. Interscience Publishers: New York.

Camougis, G. 1990. Recycling of soil contaminated with petroleum products by the AmRec process. *Hazardous Waste Management Magazine*. March:5-6.

Chapman, C.C. 1991. *Evaluation of vitrifying municipal incinerator ash.* PNL-SA-18990. Richland Wash.: Pacific Northwest Laboratory.

Chevron Chemical Co. SCRETE sulfur concrete. Manufacturers Data Sheet. San Francisco.

Colombo, P., P.D. Kalb, and M. Fuhrmann. 1983. *Waste form development program.* Annual Report. BNL-51756. Upton. N.Y.: Brookhaven National Laboratory. Sep.

Comrie, D.C., E.E. Runte, and J. Davidovits. 1988. Waste containment technology for management of uranium mill tailings. Paper presented at *117th Annual Meeting of the AIME/SME*. Phoenix.

Conner J.R. 1991. Conversation with J. Davidovits. Dec. 10.

Conner, J.R. 1985. *U.S. Patent 4,518,508*. May 21.

Conner, J.R. 1986a. *U.S. Patent 4,600,514*. July 15.

Conner, J.R. 1986b. *Method for rendering hazardous wastes less permeable and more resistant to leaching*. U.S. Patent No. 4,623,469. Nov. 18.

Conner, J.R. 1990. *Chemical Fixation and Solidification of Hazardous Wastes*. New York: Van Nostrand Reinhold.

Conner, J.R. 1992. The role of stabilization/solidification in hazardous waste stabilization. In *Proc. HazMat '92 International*. Atlantic City.

Conner, J.R. and R.S. Reber. 1992. *Australian Patent 615459*. Jan. 31.

Corum, L. 1988. Alternate Technologies. *Hazmat World*. Nov.

Cote, P.L. 1986. *Contaminant leaching from cement-based waste forms under acidic conditions*. Ph.D. Thesis. McMaster Univ. Toronto.

Davidovits, J. 1982. *Mineral polymers and methods of making them*. U.S. Patent 4,349,386.

Davidovits, J. and J.L. Sawyer. 1985. *Early high-strength mineral polymer*. U.S. Patent 4,509,985.

Davidovits, J., D.C. Comrie, J.H. Paterson, and D.J. Ritcey. 1990. Geopolymeric concretes for environmental protection. *Concrete International*. July 30.

Davis, E.L., J.S. Falcone, S.D. Boyce, and P.H. Krumrine. 1986. *Mechanisms for the fixation of heavy metals in solidified wastes using soluble silicates*. Lafayette Hill, Penn: The PQ Corp.

Donaldson, A.D., R.J. Carpenedo, and G.L. Anderson. 1992. *Melter development needs assessment for RWMC buried wastes*. EGG-WTD-9911. Idaho National Engineering Laboratory. Idaho Falls, Idaho.

DOT Canada. 1987. *Testing for solid oxidizing substances*. TP2711E. ISSN 0710-0914. Department of Transportation. Ottawa, Canada. April.

DuPont, A. 1986. *Lime treatment of liquid waste containing heavy metals, radionuclides and organics, topic no. 43*. Arlington, Va.: National Lime Assoc.

Durham, R.L. and C.R Henderson. 1984. U.S. Patent 4,460,292. July 17.

Eighmy, T.T., S.F. Bobouski, T.B. Ballestero, and M.R. Collins. 1991. *Investigation into leachate characteristics from amended and unamended combined ash and scrubber residues*. Unpublished report to the Concord. N.H. Regional Solid Waste/Resource Recovery Cooperative. University of New Hampshire. Durham, N.H.

Elnagger, H.A., A.S. Rahim, and J.G. Selmeczi. 1977. Chemical stabilization of sulfur dioxide scrubber sludges. In *Proc. Conf. Geotech. Pract. Disposal Solid Waste Matter*. ASCE, New York.

Environmental Science & Technology. 1986. 20:2.

Ezell, D. and P. Suppa. 1989. *The Soliroc Process in North America: a stabilization/solidification technology for the treatment of metal-bearing wastes with reference to extraction procedure toxicity testing*. ASTM STP 1033. Philadelphia: ASTM.

Falcone, J.S., R.W. Spencer, and E.P. Katsanis. 1983. *Chemical interactions of soluble silicates in the management of hazardous wastes.* ASTM Special Publication #851. Philadelphia: ASTM.

Franz, E.M., J.H. Heiser, and P. Colombo. 1987. *Solidification of problem wastes.* Annual progress report, BNL-52078. Upton. N.Y.: Brookhaven National Laboratory. Feb.

Franz. E.M. and P. Colombo. 1985. *Waste form evaluation program, final report.* BNL-51954. Upton. N.Y.: Brookhaven National Laboratory. Sept.

Freeman, H. 1988. *Standard handbook of hazardous waste treatment and disposal.* New York: McGraw-Hill Book Company.

Frost, D. and C. Carandang. 1990. Stabilization of organic waste. In *Proc. Superfund '91.* Washington, D.C.

Gilliam, T.M., L.R. Dole, and E.W. McDaniel. 1986. *Waste immobilization in cement-based grouts.* ASTM Special Technical Publication. Philadelphia: ASTM.

Heiser, J.H., E.M. Franz, and P. Colombo. 1989. A process of solidifying sodium nitrate waste in polyethylene. In *Environmental Aspects of Stabilization and Solidification of Hazardous and Radioactive Wastes,* ed. P. Cote and M. Gilliam. ASTM STP 1033. Philadelphia: ASTM.

Iler, R.K. 1979. *The Chemistry of Silica.* New York: Wiley.

Japan Patent. 1988. Patent number 63 35,469. February 16.

Joosten, H. 1937. U.S. Patent 2,081,541.

Kalb, P.D. 1993. Long-term durability of polyethylene of encapsulation of low-level radioactive, hazardous, and mixed wastes, 439-49. *Emerging Tech. in Haz. Waste Mgmt.* eds. D.W. Tedder and F.G. Pohland. ACS Symposium Series No. 518. American Chemical Society.

Kalb, P.D. and P. Colombo. 1984. *Polyethylene solidification of low-level wastes.* Topical Report. BNL-51867. Upton, N.Y.: Brookhaven National Laboratory. Oct.

Kalb, P.D. and P. Colombo. 1985. *Modified sulfur cement solidification of low-level wastes.* Topical Report. BNL-51923. Upton, N.Y.: Brookhaven National Laboratory. Oct.

Kalb, P.D., J.H. Heiser, and P. Colombo. 1991a. *Polyethylene encapsulation of nitrate salt wastes: waste form stability, process scale-up, and economics.* Topical report. BNL52293. Upton, N.Y.: Brookhaven National Laboratory. July.

Kalb, P.D., J.H. Heiser, and P. Colombo. 1991b. Modified sulfur cement encapsulation of mixed waste contaminated incinerator flyash. *Waste Management* 2:147-53.

Kalb, P.D., J.H. Heiser, and P. Colombo. 1992. Polyethylene encapsulation of mixed wastes: scale-up feasibility. In *Proc. Waste Management '92.* Tucson, Ariz. March 1-5.

Katsanis, E.P., P.H. Krumrine, and J.S. Falcone, Jr. 1982. Chemical reactions in an alkaline flood. In *Proc. American Chem. Soc.* Columbus, Ohio. Sept.

Koegler, S.S., R.K. Nakaoka, R.K. Farnsworth, and S.O. Bates. 1989. *Vitrification technologies for weldon spring raffinate sludges and contaminated soils phase 2 report: screening of alternatives.* PNL-7125. Richland. Wash.: Pacific Northwest Laboratory.

Kosson, D.S., H. van der Sloot, T. Holmes, and C. Wiles. 1991. Leaching properties of untreated and treated residues tested in the US EPA program for evaluation of treatment and utilization technologies for municipal waste combustor residues. In *Waste Materials Inconstruction.* Amsterdam, The Netherlands: Elsevier Science Publishers B.V.

Kulkarni, R.K. and A.B. Rosencrance. 1983. *Asphalt fixation of hazardous wastes containing heavy metal salts.* Preprint extended abstract. American Chemical Society. Washington, D.C.

Lindsay, W.J. 1979. *Chemical Equilibria in Soils.* New York: John Wiley and Sons.

Luey, J., S.S. Koegler, W.L. Kuhn, P.S. Lowery, and R.G. Wintleman. 1992. *In situ vitrification of a mixed-waste contaminated soil site: the 116-BA crib at Harford.* PNL-8281. Pacific Northwest Laboratory. Richland, Wash.

Matsushita. M. 1978. *Japan Tokyo Koho 78 42,314.*

McBee. W.C. T.A. Sullivan. and B.W. Jong. 1985. Industrial evaluation of sulfer compounds in corrosive environments. *Mining Engineering*. Jan.

McBee. W.C. T.A. Sullivan. and H.L. Fike. 1985. *Sulfur construction materials*. Bulletin 678. US Dept. of the Interior. Bureau of Mines. Washington, D.C.

McGowan, T.F. and G.R. Harmon. 1989. Transportable incineration of industrial and superfund waste. In *Proc. 10th National Conference, Superfund '89*. Washington. D.C. Nov. 27-29.

McVay, C.W., J.R. Stimmel, and S. Marchetti. 1988. *Cement waste form qualification report - WVDP Purex decontaminated supernatant. topical report*. DOE/NE/44139-49. West Valley, N.Y.: West Valley Nuclear Services Co. August.

Means, J., J. Heath, E. Barth, K. Monlux, and J. Solare. 1991. The feasibility of recycling spent hazardous sandblasting grit into asphalt concrete. In *Waste Materials in Construction*. Amsterdam, The Netherlands: Elsevier Science Publishers B.V.

MECHIM. 1975. *The Answer to Industrial Waste Problems*. Brussells.

Montgomery, D.C. Sollars, T. Sheriff, and R. Perry. 1988. Organophilic clays for the successful stabilization of problematic industrial waste. *Env. Tech. Letter* 9: 1,403.

Nehring, K.W. and S.E. Brauning. 1992. S/S process converts hazwaste for reuse. *Environmental Protection* 3:14-18, 38.

Nippon Mining Co. 1980. Japan Kokai 80 116,497. Sept. 8.

O'Hara, M.J. and M.R. Surgi. 1988. *Immobilization of lead and cadmium in solid residues from the combustion of refuse using lime and phosphate*. U.S. Patent 4,737,536. April 12.

Oberhrom, S.L. and T.R. Marrero. 1985. *Hazardous Waste and Hazardous Materials Management* 2(1): 107-12.

OWMC. 1986. Conversation with Dr. Chao. Feb. 21.

Pacific Northwest Laboratory. 1991. *Technology modules for the remedial action assessment system (RAAS)*. Richland. Wash.: Pacific Northwest Laboratory.

Payne, J.R., R.W. McManus, and F. Boelsing. 1991. *DCR treatment of oily wastes and oil-contaminated soils*. Germany.

Peacocke. 1991. *Molecular Alteration/Stabilization Technology, Third Forum on Innovative Hazardous Waste Treatment Technologies,* 37-66. EPA 540/2-91/015. Cincinnati.

Perez, J.M. and R.R. Nakaoka. 1986. Vitrification testing of simulated high-level radioactive waste at Hanford. PNL-SA-13360. Pacific Northwest Laboratory. Richland. Wash. Paper presented at *Waste Mgmt. '86 Symposium*. Tucson, Ariz. March 2-6.

Peters, J.A. 1985. *RCRA Demonstration test results of a pilot-scale glass melt furnace incineration system at Monsanto Research Corporation.* Miamisburg, Ohio: Mound Laboratory.

Petersen, R.D., A.J. Johnson, and K.G. Peter. 1986. *Nitrate salt immobilization process development and implementation.* RFP-3919. Rocky Flats Plant. Golden. Colo.: Rockwell International. Nov.

Petersen, R.D., A.J. Johnson, and S.D. Swanson. 1987. *Application of microwave energy for in-drum solidification of simulated precipitation sludge.* RFP-4148. Rocky Flats Plant. Golden. Colo.: Rockwell International. Aug.

Pojasek, R.B. 1980. *Toxic and hazardous waste disposal. Vol. 4.* Ann Arbor, Mich.: Ann Arbor Science Publishers, Inc.

Quienot, J. 1978. *Process for solidifying aqueous wastes and products thereof. U.S. Patent 4,124,405.*

Raff, R.A.V. and J.B. Allison. 1956. *Polyethylene.* New York: Interscience Publishers.

Raymont, M.E.D. 1978. *Sulfur concrete and coatings: new uses for sulfur technology of Canada Sulfur Development Institute of Canada (SUDIC).* Calgary, Alberta, Canada.

Roediger, H. 1987. Using quicklime. *Operations Forum.* April.

Ross, W.A. and C.H. Kindle. 1992. *The hybrid treatment process for mixed radioactive and hazardous waste treatment.* PNL-SA-21215. Richland. Wash.: Pacific Northwest Laboratory.

Rysman de Lockerente, S. 1976. *Treatment of wastes, especially toxic wastes, with a silicate to form a solid aggregate.* Belgian Patent 842,206.

Sandesara, M.D. 1980. *Toxic and Hazardous Waste Disposal* 4:127-33.

Santillan-Medrano, J. and J.J. Jurinak. 1975. The chemistry of lead and cadmium in soil: solid phase formation. *Soil Sci. Soc. Amer. Proc.* 39:851.

Schroeder, K., R. Graol, K.P. Ehleas, and W. Scheibitz. 1980. Ger. Offen. 2,909,572. Sept. 25.

Sell, N., M. Revall, W.Bentley, and T. McIntosh. 1992. Solidification and stabilization of phenol and chlorinated phenol contaminated soils. In *Stabilization and Solidification of Hazardous, Radioactive, and Mixed Waste.* Philadelphia: Amer. Soc. for Testing and Mat.

Simon, G.G. and W. Botzem. 1984. *Volume reduction and solidification systems for radwaste.* NUKEM GmbH, Industriestrabe 13, P.O. Box 1313, D-8755, Alzenau, Germany.

Simpson, S., H. Vidal, and M. Morris. 1988. Mixed and chelated waste test programs with bitumen solidification. Paper presented at *Spectrum '88. Nuclear and Hazardous Waste Management International Topical Meeting.* Pasco, Wash. Sep. 11-15.

Soundararajan, R., E. Barth, and J. Gibbons. 1990. Using an organophilic clay to chemically stabilize waste containing organic compounds. *Hazardous Materials Control Journal* 2(1): 42-5.

Spalding, B.P. et al. 1992. *Tracer-level radioactive pilot-scale test of in situ vitrification for the stabilization of contaminated soil sites at ORNL.* ORNS/TM-12201. Oak Ridge National Laboratory. Oak Ridge, Tenn.

SRS. 1988a. *The SRS oily sludge fixation process.* Irvine. Calif.: Separation and Recovery Systems, Inc.

SRS. 1988b. *SRS presents: the SRS fixation technology.* Irvine. Calif.: Separation and Recovery Systems, Inc.

Stanek, J. 1977. *Electric melting of glass.* New York: Elsevier Scientific Publishing Company.

Sulfur Institute. 1979. *Concrete: a new construction material comes of age, sulfur research and development.* Vol. 2. Washington. DC.: The Sulfur Institute.

Sullivan, T.A. and W.C. McBee. 1976. *Development and testing of superior sulfur concretes.* Report of Investigation 8160. Bureau of Mines. Washington, DC: US Dept. of the Interior.

Sulzer, G.G., A.A. Albert, and A.B. Palmer. 1988. Application of thermal destruction technology to the cleanup of contaminated soils. In *Proc. 5th National Conference on Hazardous Wastes and Hazardous Materials.* Las Vegas, Nev. April 19-21.

Surma, J.E., D.R. Cohn, D.L. Smatlak, P. Thomas, P.P. Woskov, C.H. Titus, J.K. Wittle, R.A. Hamilton. 1993. Graphite electrode DC arc technology development for treatment of buried wastes. *Incineration Conference '93*. Knoxville, Tenn. May 3-7.

Testa, S.M. and D.L. Patton. 1991. Paving market shows promise. *Soils*. Nov.-Dec.:9-11.

Theis, E.N. 1989. Low-density Polyethylene. *Modern Plastics Encyclopedia*, ed. R. Juran. New York: McGraw-Hill.

Timmerman, C.L. and M.E. Peterson. 1990. *Pilot-scale testing of in situ vitrification of Arnold Engineering Development Center site 10 contaminated soils*. PNL-7211. Pacific Northwest Laboratory. Richland, Wash.

Tooley, F.W. 1974. *The handbook of glass manufacture*. New York: Books For Industry. Inc.

Trezek, G.J. 1987. *Application of the polysilicate technology to heavy metal waste streams*. Sacramento, Calif.: Calif. Dept. of Health Services.

Tyco Laboratories, Inc. 1971. *Silicate treatment for acid mine drainage prevention*. Washington, DC: US EPA.

U.S. Patent 4,687,373. August 18. 1987.

US EPA. 1980. *Identification and listing of hazardous waste*. 40 CFR 261. Fed. Reg. 45. 33119. May 19. Washington, D.C.

US EPA. 1989a. *Hazcon solidification process: applications analysis report*. EPA 540/A5-89/001. Cincinnati: US EPA.

US EPA. 1989b. *Alternative Treatment Technology Information Center (ATTIC)* (data base). Washington, D.C.: US EPA, Office of Solid Waste and Emergency Response, Technology Innovation Office. Oct. 16.

US EPA. 1990a. *Alternative Treatment Technology Information Center (ATTIC)* (data base). Washington, D.C.: US EPA, Office of Solid Waste and Emergency Response, Technology Innovation Office. Oct. 12.

US EPA. 1990b. *International waste technology/geocon in situ stabilization/solidification: applications analysis report*. EPA 540/A5-89/001. Cincinnati: US EPA.

US EPA. 1990c. *Site demonstration test: soliditech solidification/stabilization process*. EPA 540/5-89/005a. Cincinnati: US EPA.

US EPA. 1990d. *Final report.* Work assignment 2-61. Contract 68 03-3413. Cincinnati, Ohio.

US EPA. 1991. *Superfund innovative technology evaluation program: technology profiles.* 4th ed. EPA 540/5-91/008. Office of Research and Development. Washington, D.C.: US EPA. Nov.

US EPA. 1992a. *Silicate Technology Corporation's solidification/stabilization process for organic and inorganic wastes in soils.* EPA 540/AR-92/010. Cincinnati: US EPA.

US EPA. 1992b. *Handbook on vitrification technology for treatment of hazardous and radioactive wastes.* EPA/625/R-92/002. Office of Research and Development. Washington, D.C.

US NRC. 1983. *Licensing requirements for land disposal of radioactive waste.* 10 CFR Part 61.

US NRC. 1991. *Technical position on waste form, revision 1. Final waste classification and waste form technical position papers.* Washington. DC: US Nuclear Regulatory Commission. Jan.

Vail, J.G. 1952. *Soluble silicates.* New York: Reinhold.

van der Sloot, H.A., O. Piepers, and A. Kok. 1984. *A standard leaching test for combustion residues.* Technical Report Bureau Energy Research Projects BEOP-31. The Netherlands.

Westsik, J.H. Jr. 1984. *Characterization of cement and bitumen waste forms containing simulated low-level waste incinerator ash.* NUREG/CR-3798. PNL-5153. Richland, Wash.: Pacific Northwest Laboratory.

Wilhite, E.L. 1987. *Concept development for saltstone and low level waste disposal.* DP-MS-86-110. Aiken, S.C.: Savannah River Laboratory. March.

Wittle, J.K., R.A. Hamilton, D.R. Cohn, P.P. Woskov, P. Thomas, J.E. Surma, and C.H. Titus. 1993. *Graphite electrode DC arc technology program for buried waste treatment.* Electro-Pyrolysis, Inc. Wayne, Penn.

Wolf, K. 1988. Solidification/stabilization and encapsulation of organic compounds from remedial actions using inorganic and organic fillers and binding agents. In *Proc. Contaminated Soils '88.* Hamburg, Germany.

Wunderly, J.M. 1956. *Waste pickle liquor disposal.* U.S. Patent 2,746,920. May 22.